ENVIRONMENTAL REMEDIATION TECHNOLOGIES, REGULATIONS AND SAFETY

Additional books and e-books in this series can be found on Nova's website under the Series tab.

ENVIRONMENTAL REMEDIATION TECHNOLOGIES, REGULATIONS AND SAFETY

THE FUTURE OF PHYTOREMEDIATION

ENVIRONMENTAL REMEDIATION TECHNOLOGIES, REGULATIONS AND SAFETY

THE FUTURE OF PHYTOREMEDIATION

JAMES C. FLORES
EDITOR

nova
science publishers
New York

NOTICE TO THE READER

Library of Congress Cataloging-in-Publication Data

Names: Flores, James C., editor.
Title: The future of phytoremediation / James C. Flores, editor.
Description: New York : Nova Science Publishers, Inc., 2021. | Series: Environmental remediation technologies, regulations and safety | Includes bibliographical references and index. |
Identifiers: LCCN 2021021022 (print) | LCCN 2021021023 (ebook) | ISBN 9781536196252 (hardcover) | ISBN 9781536196450 (adobe pdf)
Subjects: LCSH: Phytoremediation.
Classification: LCC TD192.75 .F88 2021 (print) | LCC TD192.75 (ebook) | DDC 628.4--dc23
LC record available at https://lccn.loc.gov/2021021022
LC ebook record available at https://lccn.loc.gov/2021021023

Published by Nova Science Publishers, Inc. † New York

CONTENTS

PREFACE

This book includes eight chapters about phytoremediation, which is the use of plants for the cost-effective, environmentally friendly rehabilitation of soil and groundwater contaminated by toxic metals and organic compounds. Chapter one describes which plant species are most effective for use in phytoremediation. Chapter two details how changes in the quality of light impact plant health, which in turn impacts the efficacy of phytoremediation. Chapter three presents data related to the anatomical characteristics, popular use, chemical composition, and forms of growth of Pistia stratiotes, a plant commonly used in phytoremediation. Chapter four highlights the use of constructed wetlands in the treatment and recovery of dairy wastewater. Chapter five provides the results of a study of different species of plants and their effects on the microbial activity of a limestone quarry. Chapter six summarizes the properties of purple non sulfur bacteria and their use in plant growth enhancement and bioremediation. Chapter seven examines which species of plant are most effective for removing lead and cadmium in agricultural areas. Finally, Chapter eight provides a critical view about what should be done to boost research surrounding phytoremediation in order to generate more ecofriendly restoration processes to recover polluted lands.

Chapter 1 - Hyperaccumulative plants will form the basis of phytoremediation, which will be a solution in the removal of soil pollution.

The general feature of these plants is that they are slow-growing species that produce biomass. However, the desired plant grows fast and has a good root depth and is a type that easily absorbs and absorbs the target pollutant. The phytoremediation method, which is used for the improvement of contaminated soils, has increased in recent years with the use of hyperacumulative plants. This situation will increase even more in the future. This method needs to identify plants that can adapt to different climatic conditions and different soil properties. On the other hand, it is a necessity to give more importance to the studies of determining hyperacumulative plants to answer the questions of which pollutants these plants and on which parts of the plant they accumulate. In this study, the most used hyperaccumulator families in phytoremediation method and the importance of these plants in the method in the future will be discussed. Because phytoremediation is the most economical method that can be applied to pollution removal in large areas.

Chapter 2 - Phytoremediation is gaining more and more prominence in the scientific community, as an inexpensive and easy-to-implement technique for decontamination of water, soil, and air contaminated by inorganic (heavy metals) and organic compounds (pesticides, pharmaceuticals, personal care products). Since its inception, a wide range of laboratory and field studies have been developed to evaluate different plants, to accumulate, degrade or make these contaminants less toxic. Until then, some limitations to these perspectives have emerged, such as the low capacity of the plant to remedy certain compounds that, for the most part, have stressful effects on the metabolism of the plant, decreasing its efficiency and harming the health of the plant throughout the process. Thus, it would be possible to assume that improving the plant's response to the stress caused by these contaminants would have a major impact on improving phytoremediation. However, few studies aim to improve plant health as an alternative to increasing phytoremediation efficiency, both on a laboratory scale and in the field. When dealing specifically with some organic compounds (pesticides and metabolites). Recent studies show that oxidative stress induced by pesticides alters the cellular redox balance, altering the antioxidant response as the activity of cellular defense enzymes: SOD, CAT, APX, GPx, GST, and GR (directly related to xenobiotic detoxification), in addition to levels of secondary metabolism compounds in plants, with phenolics, flavonoids, anthocyanins, and terpenes. Therefore, these parameters can be used as tolerance criteria for the selection of plants with phytoremediation capacity. One way to stimulate the production of these defense mechanisms is through changes in the quality of light, it is known that the irradiation of some colors of light can stimulate the production of secondary metabolism compounds and detoxify enzymes such as those mentioned above, in addition to this to influence photosynthetic rate, root and stem growth, and flowering. Thus, in this chapter, the authors review some of the potentials described in the literature on ways to promote phytoremediation of toxic organic compounds and maintain plant health, so that it can better cope with stressful environments and thus decontaminate more efficiently. Besides, the authors bring some unprecedented results developed in their laboratory using photo-stimulation with LEDs as a phytosanity promoter, in the face of a simulated environment contaminated with the herbicide 2,4-D.

Chapter 3 - Aquatic contamination is a worldwide problem resulting from organic and inorganic compounds of agricultural, domestic, or industrial origin. The types of contaminants found in the aquatic environment vary between heavy metals, pesticides, medicines, industrial residues, among others. Among the forms of decontamination, phytoremediation is the most convenient, due to its practicality, low costs, and being ecologically accepted. This technique uses plant metabolism to remove, detoxify, or immobilize environmental contaminants. Some plants are highly resistant to toxicity and can absorb high concentrations of different compounds from the environment and therefore stand out in the efficiency of phytoremediation. *Pistia stratiotes* has these peculiarities, and for this reason, over the years, researchers have used it to promote phytoremediation of the most diverse contaminants. This plant is a free-floating aquatic macrophyte, from tropical and subtropical ecosystems. It can be an invasive aquatic species, as it can completely cover the watercourses forming dense mats. Although this characteristic makes it inconvenient, several studies have shown that it is precisely for this reason that it can be used for phytoremediation and subsequent production of biofuel. Thus, studies with this plant date back to 1962 and are still being carried out. In this chapter, the authors will present data related to the anatomical characteristics, popular use, chemical composition, and forms of growth of *P. stratiotes*, in addition to phytoremediation processes. The toxicological and metabolic parameters related to phytoremediation with this plant will also be addressed, as well as the different studies that used *P. stratiotes* for environmental remediation in waters contaminated with heavy metals and organic compounds.

Chapter 4 - The industries seek innovative processes to treat and valorize their by-products, residues, and wastewater, considering the economic needs, the growing socio-environmental demands, and the stricter environmental legislation. The dairy industry products are responsible for relevant environmental aspects, mainly concerning energy consumption and wastewater' generation related to its life cycle. In this sense, mitigating measures should also be investigated, considering the significant number of industries and the economic and social relevance of this activity in Brazil and worldwide. The treatment and recovery of by-products, waste, and effluents within a perspective of process integration and industrial ecology are current trends. The recovery allows the cycling of nutrients and by-products, residues and wastewater in new production stages. However, the high number of small and medium industries, associated with the reduced volume of effluent generated by each company and their geographic distribution, are challenges for its application in the dairy industry, mainly related to the transport and perishability of these products. In this perspective, studies highlight the use of constructed wetlands as an alternative in the treatment and recovery of dairy wastewater. Compared to other treatment systems, the constructed wetlands have advantages such as ease of operation, a significant reduction in the environmental impact, and implementation and maintenance costs. The main component of the constructed wetlands are aquatic macrophytes, and their biomass can be used after treatment. The choice of aquatic plants must be judicious for developing the plants and maintaining the wastewater treatment. The plant's local occurrence will also indicate its adaptation to the local climate, being one of the essential factors for the development and maintenance of macrophytes during treatment. Different aquatic macrophytes are successfully used in different regions to treat domestic and industrial wastewater and provide nutrient cyclings. This chapter will address the technique of treating industrial wastewater through constructedf wetlands using macrophytes, highlighting the historical aspects, the main characteristics, and influencing factors, management, and challenges for using such technique, considering its environmental performance of nutrient cycling from the dairy industry. The results indicate that the application of constructed wetlands is promising, increasing the potential for use, both on a large and small scale, and maybe an appropriate method for application in small and medium dairy industries.

Chapter 5 - Plant species effect on microbial activity of a limestone quarry with pyrolusite's byproducts mix as cover was investigated. For the remediation project three plant species: pine, cypress, and broom were planted at 9 randomized plots. Microbial biomass (C_{mic}), basal respiration (BR), alkaline phosphatase activity (ALP), along with microbial quotient (C_{mic}/C_{org} ratio), metabolic quotient (qCO_2), and mineralization quotient (qM) were determined for a 3.5 years period from samples obtained from the rhizosphere and they used as indicators of substrate biological activity. These showed seasonal variation depending on substrate's moisture content. During the first year of phytoremediation microbial activity under broom was significantly higher than under the other two plant species (C_{mic}=225.81 μg C g^{-1}, C_{mic}/C_{org} ratio=6.04%, qCO_2=0.47 μg CO_2-C $μg^{-1}$ C_{mic} d^{-1} and qM=0.12 Σμg CO_2-C $μg^{-1}$ C_{org}). Two years after vegetation establishment, the pine treatment observed significantly higher microbial activity (C_{mic}=366.24 μg C g^{-1}, BR=120.8 μg CO_2-C g^{-1} d^{-1}, ALP=339.13 μg *p*-nitrophenol g^{-1} h^{-1}, C_{mic}/C_{org} ratio=5.85%, qCO_2=0.45 μg CO_2-C $μg^{-1}$ C_{mic} d^{-1} and qM=0.10 Σμg CO_2-C $μg^{-1}$ C_{org}) than broom and cypress treatments. However, at the same year but after a drought period with moisture as low as 10%, microbial activity under cypress plants was significantly higher than in the pine or broom rhizosphere (C_{mic}=43.49 μg C g^{-1}, BR=106.4 μg CO_2-C g^{-1} d^{-1}, C_{mic}/C_{org} ratio=1.07%, qCO_2=3.21 μg

CO_2-C μg^{-1} C_{mic} d^{-1} and qM=0.14 Σμg CO_2-C μg^{-1} C_{org}). Soil moisture, was in positive correlation with C_{mic}, ALP (r=0.31 and r=0.70, $p < 0.001$, respectively) and C_{mic}/C_{org} ratio (r=0.15, $p < 0.05$) while it was negative with BR (r=-0.19, $p < 0.01$), qCO_2 and qM (r=-0.29 and r=-0.38, $p < 0.001$, respectively).

Chapter 6 - Microorganisms are vital for agriculture as they possess biochemical reactions required for the quality and productivity of agricultural soils. Bacteria are the most diverse and intense group of microorganims and essential members of soil microorganisms as they produce phytohormones, secondary metabolites and protect plants against diseases. Bacterial diversity and amount is crucial for cycling of elements and nutrients in soil, therefore are one of the factors determining the fertility of soil. Purple non sulfur bacteria are a non taxonomic group of bacteria which are Gram negative bacteria having many growth modes. They enrich soil by fixing carbon and atmospheric nitrogen and making them available for plant uptake. This way these bacteria can be employed as biofertilizers in agriculture to enhance growth of different ground and surface plant parts. This review summarizes the properties of purple non sulfur bacteria and their use in plant growth enhancement and bioremediation.

Chapter 7 - Heavy metals pollution in agricultural lands is becoming a serious problem since it has major impact on reducing agricultural productions in quantity and quality. Meanwhile conventional efforts for remediation have not been widely implemented due to high-cost constraints. A sustainable cost-effective technology that can be applied in contaminated sites is urgently needed. Phytoremediation known as plant-base technology could be one alternative. This project aims to assess potential Pb and Cd accumulators and possible application for phytoremediation in the contaminated rice fields. Lead (Pb) and cadmium (Cd) are among the most commonly found contaminants in agricultural areas. Six wild-found plants that are predominantly grown in contaminated area were examined in an in-situ trials in Pb and Cd contaminated rice fields. Determining variables, namely biomass production, Pb and Cd uptake, Bioconcentration Factor (BCF) that indicate the ability of plants to remove metals from the soil and Translocation Factor (TF) that indicate the ability of the compound to be transferred from plant roots to shoot were examined. There are six plant species namely *Colocasia* sp. *Ipomoea fistulosa*, *Eichhornia crassipes*, *Hymenachne amplexicaulis*, *Saccharum spontaneum* and *Acorus calamus* that showed considerably high Pb and Cd accumulation after 9 weeks of growing period. The highest (BCF) for root Pb was 0,671 found in *S. spontaneum* and the highest BCF of shoot Pb was 0,558 in *I. fistulose*. Whereas the highest BCF for Cd in the root was 0,566 found in *S. spontaneum* and the highest BCF for Cd in the shoot was 0,475 in *S. spontaneum*. Translocation Factor for Pb is highest at 0,904 in *I. fistulosa* and 0,828 in *S. spontaneum*. Meanwhile the highest TF for Cd was 0,927 observed in *A. calamus* and 0,839 in *S. spontaneum*. Regarding variables of biomass production and Pb and Cd accumulation levels in the plants it can be concluded that *E. crassipes*, *H. amplexicaulis* and *A. calamus* can be suggested as potential accumulator candidates for phytoremediation of contaminated rice fields.

Chapter 8 - The Preservation of soils is a key determinant for human survival. Soil biodiversity influences the quantity and quality of food consumed, so it is a valuable natural resource. However, soil biodiversity is being affected by global man-made changes and currently, after incorrect practices in its use, there are warning signs of soil contamination with diverse pollutants. The hazardous nature of wastes can be mitigated by the application of conventional decontamination methods, such as incineration, composting and landfilling, although such techniques are expansive. A possibility is to consider a viable and promising alternative technique for treatment of contaminated soils by applying biological techniques that have as main objective the stimulation of the growth of the microorganisms using the contaminants as source of energy and food, creating a propitious environment for remediation. It is the case of Phytoremediation, a technique that uses plants and associated soil microbes to decrease the concentration or toxic effects of contaminants in environments, specially in soil. However, it is necessary to know the potential of plant around the world and, Brazil, which holds one of the greatest natural resources and biodiversity on the planet, we lack a ruge knowledge of plants with phytoremediation potential. In the last 20 years, many discussions have been raised, in both academic and business environment, converging to the same consensus: there is a delay in innovations from natural products and derivatives derived from our rich biodiversity. Therefore, this scientometric research bring a broad view of what have been carried in Brazil and abroad concerning phytoremediation studies associated with a critical view about what should be done to boos this research field in order to generate more ecofriendly restoration process to recover polluted lands.

In: The Future of Phytoremediation
Editor: James C. Flores
ISBN: 978-1-53619-625-2

Chapter 1

PHYTOREMEDIATION AND HYPERACUMULATIVE FAMILIES

Sevinç Adiloğlu* and Seda Pamay
Department of Soil Science and Plant Nutrition,
Tekirdağ Namık Kemal University, Tekirdag, Turkey

ABSTRACT

Hyperaccumulative plants will form the basis of phytoremediation, which will be a solution in the removal of soil pollution. The general feature of these plants is that they are slow-growing species that produce biomass. However, the desired plant grows fast and has a good root depth and is a type that easily absorbs and absorbs the target pollutant. The phytoremediation method, which is used for the improvement of contaminated soils, has increased in recent years with the use of hyperacumulative plants. This situation will increase even more in the future. This method needs to identify plants that can adapt to different climatic conditions and different soil properties. On the other hand, it is a necessity to give more importance to the studies of determining hyperacumulative plants to answer the questions of which pollutants these plants and on which parts of the plant they accumulate.

* Corresponding Author's E-mail: sadiloglu@hotmail.com.

In this study, the most used hyperaccumulator families in phytoremediation method and the importance of these plants in the method in the future will be discussed. Because phytoremediation is the most economical method that can be applied to pollution removal in large areas.

Keywords: phytoremediation, hyperaccumulator plants, heavy metals

INTRODUCTION

Environmental pollution has rapidly increased of late years owing to the uncontrolled use of fertilizers and synthetic pesticides and excessive industrial activities, and this has become a serious problem. All environmental sources contaminate with these toxic substances and harm the ecosystem. Various technologies have been improved to remove and reuse contaminated areas. Phytoremediation technologies are suitable, economical, and easily applicable technology for removing these toxic substances from natural environments (Tiwari et al. 2021). Phytoremediation used in the removal of various organic and inorganic complex can be carry out both *in situ* and *ex-situ*. Land degradation and the spread of pollutants through air and water are less likely in on-site applications. It also reduces the amount of waste to be stored up to 95% (Etim 2012). Although widely used to remove heavy metals, phytoremediation is also preferred for cleaning areas containing other contaminant sources such as pesticides, oils, hydrocarbons, salts, and acids. In areas where phytoremediation technologies are used, it showed many beneficial effects such as increasing soil quality through mineralization, producing biodiesel by growing biodiesel products in contaminated areas, carbon sequestration and reducing climate change effect (Tiwari et al. 2021).

Phytoremediation is often used for heavy metal pollution removal in soil, water, and air. Heavy metals divide into two as necessary and non-essential heavy metals according to the roles they undertake in the ecosystem. Plants need heavy metals like Cu, Mn, Zn, Fe, Ni to perform biochemical and physiological in the processes. However, the excessive presence of these metals makes them toxic to the plant. Metals like Pb, Cd, As, Hg on the other hand, are not essential for plants, and also, they cause highly toxic effects and environmental pollution. Also, the high amount of these heavy metals affects some biochemical and physiological staged in crop plants and reduces agricultural yield. It is important to understand the plant response to heavy metal toxicity or heavy metal accumulation to increase the cleaning efficiency of the contaminated area (Etim 2012; Adiloğlu 2016; Adiloğlu; 2017; Adiloğlu and Göker 2021; Adiloğlu 2021).

In this section, a review has been made by explaining the giving brief information about the hyperaccumulator plants that have been used in phytoremediation studies for heavy metal pollution in recent years and/or which have been declared suitable for phytoremediation of heavy metal together with the families to which these plants belong.

MECHANISM OF PHYTOREMEDIATION

The ecological method in which plants are used, phytoremediation is used to remove contaminants that contaminate environmental parameters (sludge, soil, sediment, groundwater, etc.) limit their movement and degrade. The use of plants in these contaminated areas also seems aesthetically pleasing. Plants use sunlight to remove pollution and do not need extra chemicals, making this method an environmentally friendly method. With the phytoremediation method, metals, crude oil, pesticides, polycyclic aromatic hydrocarbons, solvents, and wastewater can be removed from natural environments. There are different methods in phytoremediation applied according to the working environment and plant (EPA 1999). These methods are briefly described in Table 1. Plants react differently at heavy metal concentrations depending on whether they are toxic or not. Many plants are sensitive to these heavy metals nor yet at low concentrations, but some plants show resistance to the toxicity of heavy metals. These plants are called hyperaccumulator plants and these plants use in phytoremediation technology (Barceló and Poschenrieder 2003).

Table 1. Subclasses of phytoremediation (EPA 1999; Barceló and Poschenrieder 2003; Etim 2012)

Method	Features
Phytoextraction	The most common subclass of phytoremediation, the phytoextraction method involves taking the pollutant from the subsurface and moving it to the shoot. With this method, it is aimed to reduce the heavy metal density in the environment by growing plants with high biomass for metal accumulation in the shoots. In phytoextraction, plants capable of high metal accumulation and high biomass are preferred.
Rhizofiltration	This technique is based on the elimination of heavy metals by plant roots from surface water and wastewater by precipitation or absorption. The plants used in this method should not only be resistant to heavy metal concentrations but also the plant roots should have high absorption ability. After the heavy metal accumulation in underground part of the plant, the harvested plants are burned or composted in order not to mix the pollutants back into the environment.
Phyto-stabilisation	In this method, the plant restricts the movement of heavy metal pollutants, prevents them from entering groundwater and entering the food chain by reducing the bioavailability. This technique can be used to create a vegetative cover to solve the heavy metal concentrations occurring in the surface soils and the deficiencies in the vegetation on the surface. Thus, soil loss that may occur due to precipitation and wind is prevented, while the contamination of the soil with groundwater can be prevented.
Phytodegradation	In this method, organic pollutants are degraded with enzymes etc. that secreted by the plant. These pollutants are broken down and degraded by the plant. They are then taken into plant tissues by plants and used as food.
Rhizodegradation	In this method, organic pollutants are decomposed by soil microorganisms and they consume these substances for needs such as energy and nutrition. Natural substances secreted by plant roots help microorganisms in this process and increase microbial activity. In addition, plants loosen the soil and carry water with their roots accelerates the rhizodegradation.
Phytovolatilization	In this method, organic pollutants taken by plants are released into the atmosphere consequently the plant's transpiration. The pollutant released into the atmosphere can be in its own form or in a form modified by the plant.

HYPERACCUMULATOR FAMILIES

Brassicaceae Family

Brassicaceae or cabbage family (such as *Arabidopsis thaliana*, *Raphanus sativus*, *Brassica oleracea*, *Brassica napus*, *Armoracia rusticana*, *Brassica rapa* and others) (Figure 1) are well-recognized metal accumulators. High metal translocation in plants has made plants for the remediation of these different metal polluted sites. These plants are processed by separating the upper parts of heavy metals into the spaces of the above-ground parts, especially in the leaves. Studies are ongoing to understand the mechanism and diversity of omics in the Brassicaceae family of plants related to conduction hyperaccumulation and hyper-tolerance properties, metal aggregation and detoxification (Daud et al. 2018).

Figure 1. Brassica species (Williams 2015).

Figure 2. *Brassica oleracea* var. Acephala (Akulova 2018).

Plants used as hyperacumulators have high metal content and low biomass. This is an economical and applicable method for both safe disposals of non-recoverable waste and residues of metals. Besides this, non-accumulator plants create a considerable amount of biomass with poor metal accumulation, which is not economical to treat metals for recovery and is also disposed of in some way.

Good results are obtained in a study using *Brassica juncea* (Indian mustard) to clean soil contaminated with lead (Pb). Some research has shown that *Brassica juncea* much better accumulates the Pb heavy metal in the roots, then deposits it in the fruit and branches as well. The higher Pb concentration in the root is owing to the depth of the roots in the soil and the ability to accumulate roots for lead. Lead accumulation in soil and plant parts of Brassica juncea plant was examined. In this study, While the effect of lead pollution is highest between plant roots and soil, it has been determined that the lowest lead value is between uncontaminated soil and shoots (Tatu, Vladut, Iulian Voice, Vanghele, & Pruteanu, 2020).

Ornamental cabbage (*Brassica oleracea* sp. Acephala) (Figure 2) is a plant that can be grown mostly from early autumn to late winter. Unlike most of the other plants used for phytoremediation, the season range in which this plant is grown is different, supporting that cabbage can be a suitable option for phytoremediation and can be used in the urban landscape in autumn and winter, particularly in metropolitan areas with high lead (Pb) contamination. Kafi et al. has been reported that cabbage plant can generally absorb and accumulate lead under salinity stress (Haghighi, Kafi, Pessarakli, Sheibanirad, & Sharifinia, 2016).

Brassica napus, *Brassica juncea* and *Raphanus sativus* belonging to the Brassica family are recommended as heavy metal accumulators for plant breeding studies. Overall results from the study of Marchiol et al. exhibited a positive nonlinear relation between their numbers in the soil and the heavy metal accumulated in the roots and shoots of radish (*Raphanus sativus*). It has also been observed that when it is consumed in the soil, its accumulation in plant tissues increases. It has been observed that the main lead accumulation occurs in roots instead of shoots (Marchiol, Assolari, Sacco, & Zerbi, 2004).

Hamvumba et al. has observed that *Brassica rapa* var *chinensis* (Chinese cabbage) can be accumulated the Pb in roots at all levels of Pb. As the Pb contamination in the soil increased, there was a regular increase in the roots of *Brassica chinensis* (Figure 3) and this result showed that can be used for the Pb phytoremediation of the plant (Hamvumba, Mataa, & Mweetwa, 2014).

Brassica juncea shown in Figure 4 has a significant effect on zinc (Zn) accumulation. Considering the accumulator capacity, it was realized that the maximum value for Zn accumulation in the plant was in the stem and the minimum value in the roots. It has been reported that this plant can accumulate extraordinary amounts of Zn concentrations in overground parts (Tatu et al. 2020).

In a study by Raj et al. when the *Brassica juncea* plant was tested for the phytoremediation capacity of mercury (Hg) from the soil, it was observed that this plant had a high Hg accumulation capacity. Phytoremediation potential has been noted as 2.62 mg Hg kg $^{-1}$ plant $^{-1}$ (Raj, Kumar, & Maiti, 2020).

Figure 3. *Brassica rapa* L. subsp. *chinensis* L. (Original).

To analyze the tolerance threshold and aggregation of *Brassica juncea* L. for Pb, experiment was established under greenhouse terms using different doses (80-136-362-1150 mgkg^{-1}). Plant height decreased from

1.37 to 0.83 m with Pb contamination in the soil and soil Pb elevation increased to 1150 $mgkg^{-1}$. With the increase of lead concentration in the soil, the amount of lead in shoots and roots increased. Consequently, this study, it was observed that *B. juncea* can accumulate Pb in the soil in summer (Bassegio et al. 2019).

Figure 4. *Brassica juncea* (NC State University 2020).

Cardamine violifolia plant belonging to the *Brassicaceae* family is the first selenium hyperaccumulator to be discovered which has belonged to *Cardamine* class about selenium accumulation, that is, high selenomethionine abundance. It is predicted that this plant can be used for the removal of Se pollution in the high rate of soil and water resources in China (section; up to 2.4 mg and up to 275μg water-soluble Se / kg soil and Se / L water) (Both et al. 2020).

In the study investigates the accumulation, tolerance, and toxicity of zinc (Zn) in Indian Mustard (*Brassica juncea*), it has been proven that the tolerance index of the plant increases with increasing metal concentration. The translocation factor for each Zn concentration has been noting to be greater than 1, indicating that a greater amount of metal was deposited from the roots of the aerial parts of the plants. Zn deposits have been found in the epidermis, vascular bundle, and mesophyll cells of plant leaves. The plant has evolved by tolerating even higher Zn concentrations. Zn accumulation in leaf tissues also increased with increasing Zn concentration (Chaudhry et al. 2020). In the study realized by Małecka et al. with *Brassica juncea* (Indian mustard), the plant was exposed to different contaminator (cadmium, copper, lead, and zinc) for 96 hours with a concentration of 50 μM. Then, it was harvested for analysis. A tolerance index (IT) higher than 90% was observed for all *B. juncea* plants (Małecka et al. 2019). *Brassica campestris has* shown in Figure 5, which belong to the *Brassicaceae* species, has been proven by the studies that it has potential use for the phytoextraction of Zn, Cd and Cu heavy metals. *Sinapis arvensis* (Figure 6) and *Thlaspi arvense* plants are also members of the *Brassicaceae* family and are suitable for use in the removal of Zn and Pb pollution in the soil. *Rorippa palustris* plant is also used to remove the pollution caused by Zn and Cd heavy metals (Drozdova et al. 2019).

Figure 5. *Brassica campestris* (Chen et al. 2019).

Figure 6. *Sinapis arvensis* (Near East University, 2020).

Radish (*Raphanus sativus* L.), another plant belonging to the *Brassicaceae* family, is an edible root vegetable. According to the results obtained from the variance analysis performed after selenium (Se) applications in radish plants, the the dry weight of the plant's root and stem were not affected by Se application. However, the Se content in leaves and roots was influenced by the binary interaction with application forms against selenium sources. When the Se was applied as selenate (SeO_4^{2-}) in soil, the highest accumulation of Se was observed in radish plants (Figure 7) leaves and roots (da Silva et al. 2020).

Figure 7. *Raphanus sativus* L. (Yüzüncü Yıl University, 2020).

In a research conducted on a metal-polluted soil in a city, it was shown that *A. halleri*' have the potential to a certain extent removes Zn and Cd heavy metals from the soil. *A. halleri*'can grows in alkaline soil, and this plant was able to accumulate Zn and Cd in its above-ground parts, although soil conditions were not suitable for phytoremediation. These consequences highlight the possibility of using *A. halleri*'in on multifarious soil types in phytoremediation studies (Grignet et al. 2019).

In the field study carried out in Tekirdağ, the chromium (Cr) removal capacity of canola (*Brassica napus* L.) from agricultural areas were investigated, each parcel has Cr $(NO_3)_3$ as chromium (100 mg kg^{-1}) and more. EDTA has performed in 0, 5, 10, 15 mmol kg^{-1} doses pending the flowering stage of the plants. It was observed that the root and shoot efficiency of the *Brassica napus* L. (Figure 8) harvested after the four-month trial period decreased with increasing EDTA doses; Cr content of plant stem and roots increased. With these results, which are statistically significant at the 1% level, it has been shown that Cr can be clear of the contaminated soil by phytoremediation method in Tekirdağ province lands (Adiloglu and Turgut Saglam Ali Süme 2015).

A field test research was done to clear up Nickel contamination in agricultural soils using canola (*Brassica napus* L.). Nickel (100 mg kg^{-1}) was applied to each plot as $NiSO_4$ forms, and four doses of EDTA (0, 5, 10, 15 mmol kg^{-1}) chelate were applied to each plot during the flowering period of the plants. According to the field test results of the plants harvested four months after planting, while the root and shoot efficiency of the *Brassica napus* L. reduced with increasing EDTA chelate doses. The amount of Ni in the roots and shoots of the canola enhanced with the increase of EDTA. According to the statistically significant results at the level of 1% obtained from the testing, it has been observed that the soil pollution of Tekirdağ can be removed from the soil by the plant (Adiloğlu, Turgut Sağlam, et al. 2016).

Figure 8. Canola (*Brassica napus* L.) (Original).

Canola (Brassica napus L.) used as a plant for the removal of cobalt (Co) pollution, cobalt in the form of $CoS0_4$ (100 mg kg S^{-1}) and EDTA chelate (0, 5, 10, and 15 mmol kg $^{-1}$) was applied in each pot. Four months after planting the plants were harvested. Though the amount of Cu, Fe, Mn present in the soil sample taken from the experimental area was sufficient, it was determined that the Zn amount was insufficient. As a result of the experiment, it was observed that while increasing EDTA doses harmed the above ground and underground components of canola, the Co concentrations in the root and stem of the plant increased statistically by 1%. The result shows that Co heavy metal pollutants in Tekirdağ province soil can be improved by phytoremediation method using *Brassica napus* L. (Adiloğlu 2016).

Asteraceae Family

The *Compositae* (*Asteraceae*) family is one of the largest flowering plant families, comprising over 23,600 known species, 1,620 genera and 13 subfamilies. It is often used in traditional medicine for infections, as a microbial treatment, and as a spice (Z. Liu, Chen, and He 2018). They are also widely used in landscaping due to their cosmopolitan distribution and ecological adaptation (Nikolić and Stevović 2015). Especially *Taraxacum, Tagetes, Zinnia, Centaurea* and *Gerbera* species belonging to this family are the species widely used in landscape greening and garden decoration (Z. Liu, Chen, and He 2018). One of the most significant qualities of this family is that there are big differences between members. The plants belonging to this family range from perennials to shrubs, vines or trees (Nikolić and Stevović 2015). In a study, some plants belonging to *Asteraceae* family (*Taraxacum mongolicum* Hand. -Mazz., *Zinnia elegans* Jacq., *Tagetes patula* L., *Centaurea cyanus* L., and *Tagetes erecta* L., *Gerbera jamesonii* Bolus) were found in soil samples containing different concentrations of Cd heavy metal. hyperaccumulator properties were examined. Although the Cd concentration of *Tagetes patula* L. and *Tagetes erecta* L. plants (Figure 9) increased, their development was the same as the control plant. However, it was observed that the growth of other plants slowed down, and their height was shorter. This study shows that *Tagetes patula* L. and *Tagetes erecta* L. can be used for removal of Cd in soil (Z. Liu, Chen, and He 2018).

It has been reported that the *Baccharis trimera* plant (Figure 10) relating to the *Asteraceae* family has the highest copper concentration (586.2 $mgkg^{-1}$) compared to other species that grow spontaneously in copper mining wastewater. In the same study, this herb was reported to be a hyperaccumulator in vanadium and barium metals. Regarding lead levels, both *Baccharis dracunculifolia* (Figure 11) and *Baccharis trimera* have accumulated at high concentrations (Silveira Rabelo and Caldeira Costa 2018).

Atractylis cerratuloides (Figure 12) has been performed in many studies to evaluate its hyper-accumulation potential. This type of fluorspar belonging to the *Asteraceae* family was used in an area contaminated with mine waste. It has been determined that copper and manganese accumulate in the root of the plant (Bayouli et al. 2020).

Figure 9. *Tagetes patula* L. (Z. Liu, Chen, and He 2018) *Tagetes erecta* L. (Z. Liu, Chen, and He 2018).

Figure 10. *Baccharis trimera.* (Sırasıyla; çalı form, slindirik yapraksız dallar veya sapsız yapraklar, ezilmiş bitki örneği) (Silveira Rabelo and Caldeira Costa 2018).

Figure 11. *Baccharis dracunculifolia* ssp. tandilensis (Echeverría, Alonso, and Comparatore 2017).

Figure 12. *Atractylis serratuloides* (Bakali 2016).

In another study, *Baccharis neglecta* Britt. (As) (Figure 13), *Bidens humilis* (As), *Isocoma veneta* Greene (Cd, Ag, As, Pb, Cu, Zn), and *Viguiera linearis* Sch. (Cd, Cu, Mn, Pb, Zn) plants have been reported to be used as hyperaccumulators for different heavy metals (Kaur, 2020).

The researchers are investigated BCF (Bioaccumulation factor) and TF (Translocation factor) of 16 plants in the *Asteraceae* and evaluated their phytoremediation potential. In the study, Cd accumulation was determined for *Taraxacum officinale* F. H. Wigg (162 $mgkg^{-1}$) and *Crepis sp.* (13 $mgkg^{-1}$) plants that shown in Figure 14. The maximum BCF of Pb and Cd was found to be as 0.326 and 1.787, respectively (Hesami, Salimi, and Ghaderian 2018).

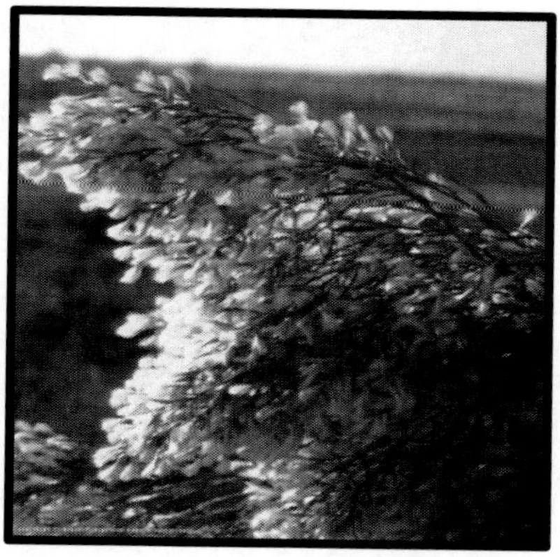

Figure 13. *Baccharis neglecta* Britt (Texas A&M University, 2020).

Sonchus asper (L.) Hill (prickly thistle), shown in Figure 15. In a study realized in the red mud storage area, the utilization capacity of this plant in phytoremediation for Fe, Mn, Cu, Zn, Cd, Cr, Pb was tested. It was follow up that the concentration of heavy metals in Fe (2363.5 μgg^{-1}) and Cr (66.50 μgg^{-1}) in the plant was higher than the concentration in the soil (Mishra and Pandey 2019).

Launaea asplenifolia L. is a widespread plant worldwide but it is more common in tropical regions. In the study carried out in the red mud storage area containing different heavy metals, it was observed that this plant only accumulates Fe heavy metal (2931.65 $\mu g\ g^{-1}$ dry weight (Mishra and Pandey 2019).

Figure 14. *Taraxacum officinale* F. H. Wigg *Crepis sancta* subsp. obovata (photo B. Biel) (Gómez-Arroyo et al. 2018) (Vassilev and Pedashenko 2013).

Figure 15. *Sonchus asper* (L.) Hill (Drobot, 2020).

Figure 16. *Cirsium vulgare* L. plant (Orijinal).

A study was carried out on the removal of Cr from polluted soils by using *Cirsium vulgare* L. (Figure 16) with phytoremediation method. 30 $mgkg^{-1}$ Cr heavy metal in the form of CrO_3 was applied to the pots as a contaminant, together with EDTA in doses of 0, 3, 6, 8 and 10 mmol kg^{-1} to facilitate the intake of the plant. After the two-month trial period, it was observed that the increasing EDTA dose increased the intake of other plant nutrients (Fe, Cu, Zn and Mn) in Cr applied pots to a certain extent. According to the results obtained in this study; *Cirsium vulgare* L. plant emerges as an effective accumulator plant that can be used for phytoremediation of soils polluted with chromium (Dökmeci and Adiloğlu 2020).

Lamiaceae Family

The *Lamiaceae* (Labiatae) family has a very large distribution, comprising about 236 genera and between 6900 and 7200 species. Representatives of the *Lamiaceae* family are widely used for various purposes, but their most common use is in the food industry, as they are characterized by a rich content of aromatic compounds. Sweet basil, mint, rosemary, sage and thyme are just a few of the types commonly used in culinary production all over the world (Fidan et al. 2019). Lavender (*Lavandula vera* L.) was raised for phytoremediation of Pb, Cd, and Zn polluted areas. *Lavandula vera* L. is accepted as a Pb hyperaccumulator. It is thought that the cost of plant breeding will be significantly reduced with lavender grown in heavy metal contaminated soils (Pandey, Rai, and Korstad 2019). In a study on the removal of nickel by phytoremediation, five different Ni doses (0, 5, 10, 20, 40 mgL^{-1}) were applied to Lavender (*Lavandula angustifolia*) and Mojito (*Mentha × villosa*) plants (Figure 17) grown in greenhouse conditions. As the accumulation increased as a result of the increased Ni ratios applied, no apparent toxic or inhibitory signs were observed in Lavender. It was concluded that these plants can be grown for Ni phytoremediation under favourable conditions (Barouchas et al. 2019).

Studies show that aromatic and medicinal plants can be growable successfully and may have a potential for phytoremediation in heavy metals contaminated environments. *Ocimum basilicum* L. and *Melissa Officinalis* L. plants (Figure 18) are classified as accumulating Pb (lead), Cd (cadmium) and Zn (zinc). Also, *Lavandula spica* L. has been reported as a lead, zinc and cadmium accumulator and *Nepeta cataria* L. as a zinc accumulator. It has been observed that *Salvia sclarea* L. can accumulate Pb, Zn and Cd in extremely high concentrations. The high accumulation of Pb by *Salvia sclarea* L. (Figure 19) indicates that it is a hyperaccumulator for this heavy metal and has the potential to be used in clearing the contaminated soils with Pb (V. Angelova 2013).

A pot experiment was conducted to investigate whether the yield and chemical composition of essential oil contained in *Mentha crispa* L. were affected by increasing application (900, 1,800, 3,600, 7,200, 9,000 $mgkg^{-1}$) of lead.

Figure 17. *Lavandula angustifolia Mentha × villosa* (BitkiVT Online Bitki Veritabanı-İTU, 2020; University of Wisconsin Madison, 2020).

Figure 18. *Ocimum basilicum* L. (Li and Chang 2016) *Melissa Officinalis* L. (Abdel-Naime et al. 2016).

M. crispa is a plant that accumulates and tolerates high Pb concentrations, it is not be considered of the Pb hyperaccumulator. Because, considering the essential oil production, the high level of Pb ratio significantly affects the essential oil content and yield in the plant.

Therefore, under conditions of extreme contamination, *M. crispa* is not considered being the Pb hyperaccumulator (Sá et al. 2015).

Figure 19. *Salvia sclarea* L. (Nadda et al. 2012).

A field study was performed in a metal-contaminated area to commentate the usability of *Salvia sclarea* L. in phytoremediation of polluted soils. The heavy metal content of *Salvia sclarea* L. in root, leaf, stem, and inflorescences was determined by ICP instrument. As a result of the work done, it has been established that the *Salvia sclarea* L. accumulates heavy metals by the way of the plants' root system. However, extraordinarily little of the heavy metals were trapped by the roots and accumulates heavy metals in its aboveground parts (shoot, leaves, inflorescences). The study has shown that this herb is tolerant of heavy metals and is an accumulator for Cd and Zn while a hyper-accumulator for Pb heavy metal. The Salvia sclera L. can be used successfully in heavy metal contaminated soils with phytoremediation method (V. R. Angelova et al. 2017).

In one study have compared to As (III) accumulation in three different species of *Ocimum* and the effects observed in each plant after exposure. Plants have incubated for 10 days at a concentration of 100 μM As (III) and then examined. A high amount of As accumulation was observed in the roots: 662 $\mu g\ g^{-1}$ in *O. tenuiflorum*, 764 $\mu g\ g^{-1}$ in *O. basilicum* and 831 $\mu g\ g^{-1}$ in *O. gratissimum*; dry weight (DW). Arsenic (III) is none of the plants tested was found in the oil content. This study shows that *Ocimum* species can be preferred in phytoremediation under favourable conditions and its essential oil can also be used (Siddiqui et al. 2013).

Lavender (*Lavandula vera* L.) is a plant that is heavy metals and can accumulate heavy metals in both the root and aboveground parts. However, only tiny amounts of heavy metals accumulate in the roots, and most of them accumulate in the stems, leaves and branches. In a study conducted with *Lavandula vera* L. in soils pllution lead, Zinc and Cadmium heavy metals in Bulgaria, the values of the plant have been obtained in the roots, respectively: 1566.9 $mgkg^{-1}$, 1755.9 $mgkg^{-1}$ and 160.9 $mgkg^{-1}$. These values that have been obtained in the roots are higher than the values considered toxic for plants (0.1 mg kg^{-1} Cd, 30 mg kg^{-1} Pb, 100 mg kg^{-1} Zn). As a result of the study, the data have shown that *Lavandula vera* L. plant can accumulate heavy metals in its roots and can tolerate pollution when grown in areas with heavy metal pollution. (Violina R. Angelova et al. 2015).

The metabolism and biosynthesis of phytochemical components in medicinal and aromatic plants are significantly moved by ecological conditions. In the study, growth, and nutritional behaviours of pennyroyal (*Mentha pulegium* L.) at different Cu (0.5, 25 $mgkg^{-1}$) - Zn (0, 10, 50 $mgkg^{-1}$) concentrations and also, essential oil content has examined. It was made to investigate variations in content and composition. The quality of the constituent components and quality of *M. pulegium* (Figure 20) essential oils, the volatile components of pulegone, sabinene, 1,8-cineol, cis-isopulegone, a-pinene and thymol, was considerably increased with copper and zinc application in comparison to the control plant. In this respect, pennyroyal with rich content can be used as an accumulative plant for agricultural soils polluted with Cu- Zn heavy metals (Lajayer et al. 2017).

Two culture trials, under greenhouse conditions, conducted to investigate the phytoremediation capacity of *Rosmarinus officinalis* L. (Figure 21) in soils polluted with Cd-Zn by Aram and colleagues. Pollutants have applied in three replications, as six Cd doses (0, 10, 20, 40, 80 and 160 ppm) and six Zn doses (0, 25, 50, 100, 200 and 400 ppm).

When the results were examined, it was observed that heavy metal concentrations in roots and shoots increased with increasing doses. The highest TF (translocation factor) value for Cd and Zn was determined at 80 and 25 ppm doses and the maximum BF (bioaccumulation factor) value was observed in plants with 80 and 400 ppm doses, respectively. Trial results have shown that rosemary can accumulate heavy metals of Cd and Zn on their branches and be used for phytoremediation in contaminated soil (Aram, Negahdarı, and Tajgardoon 2019).

Figure 20. *Mentha pulegium* L. (Morse 2016).

Figure 21. *Rosmarinus officinalis* L. (Plant Biodiversity Conservatory and Research Center 2020) (Carrubba, Catalano, & National, 2009).

Fabaceae Family

The *Fabaceae* family (Figure 22) is a large family of flowering plants known as the legume family, pea family or bean family and is economically important. The *Leguminosae* family, older name than the still considered valid Fabaceae, represents another plant family of large flowering plants after *Orchidaceae, Fabaceae* and *Asteraceae*. Also, the *Leguminosae* family comprising 800 genera and 20,000 species, according to the Royal Botanic Gardens (Nofel 2016; Bianucci et al. 2020). Members of this family are the largest genus is *Astragalus* (more than 2,000 species), *Acacia* (more than 900 species), *Indigofera* (about 700 species), *Crotalaria* (600 species), and *Mimosa* (500 species). Besides, the *Vigna* genus is a member of the *Leguminosae* family and uses for many traditional purposes (Nofel 2016). In nature, these plant family play central roles as a source of protein for humans and other living things. These plants increase the amount of nitrogen, phosphorus, and humus in soil content, as well as reduce greenhouse gas emissions (Bianucci et al. 2020).

Vicia spp is a preferred plant with both its agricultural advantages and phytoremediation capabilities. It has been reported as a result of studies that *Vicia* species can deposit metal/metalloid in their roots. *V. faba* accumulated As, Cd and Cs in its roots, while *V. villosa* accumulated As in its roots when grown in contaminated areas. It has been reported that mostly Pb, Zn and Cd heavy metals accumulate in the above-ground parts of *V.sativa* and *V. faba* (Ibañez, Medina, and Agostini 2020).

Mung bean (Figure 23), a rich source of protein commonly found in subtropical and tropical Asia, is a plant belonging to the family Fabaceae. Bianucci et al. based on their study report with mung beans, they stated that *Vigna radiata* seedlings can be used as bioindicators and detect the level of As pollution in nature (Bianucci et al. 2020).

In a study in which the bioaccumulation capacity of *Vachellia campechiana* plant (Figure 24) and its morphological changes as a result of pollution were tested under greenhouse conditions, it was observed that Pb, Fe, Cr, Cu and Zn heavy metals accumulated in the roots and leaves, respectively. As a result, *V. campechiana* is a plant that has the potential to be used in phytoremediation with its ability to protect itself in contaminated areas, it's capacity to bioaccumulate metals that it does not need in roots and leaves, and its high heavy metal translocation rate (Santoyo-Martínez et al. 2020).

Figure 22. Fabaceae family (Harvard Health Publishing 2018).

Figure 23. Mung bean seed and Mug been (Ganesan and Xu 2018).

Hydroponic experiments were carried out with *Pongamia pinnata* L. seedlings to research into the effect of arsenic (As) concentrations. As a result of the experiment, it was observed that the plant could tolerate up to 0.2-1.0 mM As concentration and did not indicate a noticeable reduction in chlorophyll content and growth. Also, chlorophyll content and growth values were reduced at 1.5 mM As concentration. Accumulation of approximately 1129-3322 mg kg^{-1} (DW) in the roots of *P. pinnata* samples grown at 0.2-1.5 mM As concentration and 345-3662 $mgkg^{-1}$ (DW) on shoots were observed. Consequently, the bioaccumulation and translocation factor values have been observed greater than 1. And these values have confirmed that this plant can accumulate higher amounts of As (D. Kumar et al. 2017).

Figure 24. *Vachellia campechiana* (Australian National Botanic Gardens 2016).

The study was regulated to evaluate the phytoremediation (Figure 25) of *Miscanthus sinensis* belonging to the family *Fabaceae* at a soil mercury (Hg) concentration between 1.48 and 706 mg kg^{-1} and to investigate the effects of mercury on soil microbial community and nutrients. The changes in biomass yield values in SOD activity, dry mass obtained as a result of the experiment showed that this plant was able to tolerate higher Hg concentrations and could continue to grow even at Hg concentrations until 706 mg kg^{-1} in soil. This study has proved that *Miscanthus sinensis* is a potential accumulator plant that can be used to improve mercury-contaminated soils. It has been stated that *Miscanthus sinensis* is a promising energy plant for phytoremediation as well as biofuel production in areas contaminated with Hg (Zhao et al. 2019).

Figure 25. *Miscanthus sinensis* (Texas Tech University, n.d.).

Figure 26. *Prosopis juliflora* (Gallaher and Merlin 2010).

The growth and development of *Pongamia pinnata* L. were observed in 20- and 40-mL L^{-1} wastewater discharged together with wastewater discharged from a local tannery. As a result of the analysis, the obtained physicochemical values were higher than the physicochemical properties evaluated for wastewater quality and higher Cr concentration was detected compared to other metals. Regarding plant growth and pigments species, *P. pinnata* had decrease ($p \leq 0.05$) in root length, dry shoot biomass, chlorophyll, leaf area, and carotenoid amount. However, the plant collar diameter, shoot length, and root dry weight was not affected by pollution. It was observed that the roots of plant accumulate 6 and 11 times more chromium at compared to the control in 20- and 40-mL L^{-1} wastewater treatment, respectively. The study has been shown that *P. pinnata* has the potential to act as phytoremediation both as a solution to water scarcity by cleaning contaminated water and to alleviate metal contamination (Abid et al. 2019).

Extensive research has been done. Heavy metals of Cd-Pb have been found in the leaves, roots and soils of plants found on the roadside. The translocation factor of Cd in *Pongamia pinnata* L. plant that used in the study was determined as 1.30 ± 0.50. According to the researched roadside soil data, for Cd, it was concluded that *P. pinnata* L. plant has high translocation factor and is a suitable candidate for phytoextraction technology for the clean up of this toxic element from the soil (Rai, Sharma, and Panchal 2019).

Prosopis juliflora that has shown in Figure 26, family *Fabaceae*, is one of the common examples of plants be able to accumulate heavy metals in high concentrations in their shoots. It has been reported that it can be used to remove harmful factors such as chloride, SO_4, Cr^{+6}, NO_3 etc. extracts in leather industry wastewater. *P. juliflora* is known to have high metal deposition rates under natural conditions (Keeran et al. 2019).

Poaceae Family

The *Poaceae* family, one of them the largest angiosperm families, spreads from the polar circles to the equator and from mountain tops to the sea. It is generally ecologically dominant and is estimated to contain 25% of Earth's vegetation. This family consists of about 11,000 species and 700 genera and six subfamilies. Many cultivated plants such as wheat, rice and corn are included in this family. While some species such as bamboo are used for structural purposes, many species are considered as weeds (Khan et al. 2017).

It is known that *Puccinellia distans* (Figure 27), which is an alkaline grass, shows hyperaccumulation properties against excessive boron concentrations and tolerates high salinity levels. Therefore, Se tolerance capacity and accumulation of this plant were investigated. *P. distans* has been found to accumulate 1500 - 2500 times more Se when grown in a hydroponic condition containing 80 or 120 μM Se compared to plants found in an element-free environment. As a result of this study, it has been discovered that *P. distans* is a new Se accumulator and can be used in phytoremediation (Kök et al. 2020).

In a study performed to research the behaviour of *Lolium perenne* L. plant in an environment with Pb ions. The soils containing Pb at 250, 500, 750, 1000 mg kg^{-1} concentrations were used with the control soil. Consequently, statistically results in all differences between various lead concentrations in germination, vegetative and physiological characteristics. Therefore, it has been predicted that *Lolium perenne* L. plant can be used in soils contaminated, up to a concentration of 1000 $mgkg^{-}1$, with Pb heavy metal (Gholinejad et al. 2020).

Salas-Moreno & Marrugo-Negrete have been investigated the growth behaviour of *Paspalum fasciculatum* Willd ex Flüggé in gold mine soils contaminated with Cd and Pb. Three different concentrations have used for Cd and Pb: 15, 30 and 50 mg kg^{-1}. After 90 days of incubation of the plant, the ability to accumulate and assimilate cadmium and lead, translocation and bioaccumulation factors have evaluated. At plant tissues, Cd-Pb heavy metals have shown the same behaviours with decreased amounts of Cd and Pb accumulation in roots > leaves > stems. It has been observed that *P. fasciculatum* has a stabilizing effect in terms of high metal concentrations and less translocation in tissues for Cd while showing phytoextraction capacity for Pb. Besides, the plant is thought to increase the amount of organic matter and pH in the soil (Salas-Moreno and Marrugo-Negrete 2020).

Vetiver grass (Figure 28), which belongs to the family *Poaceae*, is a perennial grass with a short rhizome and a large, slender root system. The deep root structure of the plant prevents it from dislodging in areas with strong water flow. Vetiver grass (*Cymbopogon citratus*) has resistance to different environmental impact and high-quality ecological adaptability. Also, its ability to grow in dry and rocky areas and to adapt to chemically contaminated areas are other important features that make the plant stand out. These properties enabled it to be used for the removal of the most common heavy metals such as Co, Zn, Ni and Pb emitted to the environment in mine waste channels. This plant species has been deemed suitable for the removal of heavy metals pollution from mine drainages and mine waste found in Ethiopian mining sites (Gebru 2019).

Figure 27. *Puccinellia distans* (Matson 2011) *Puccinellia distans* (Flora of Eastern Washington and Adjacent Idaho, 2014).

Figure 28. Vetiver grass *(Cymbopogon citratus)* (Gebru, 2019; Torres-Avilez, Silva, Albuquerque, & Paulino, 2007).

Generally, water resources contaminated with lead are an environmental problem that exists in the world. Pb is a heavy metal with high toxicity and therefore the determination of the accumulator plant for Pb is very important for the reclamation of contaminated areas. For this purpose, phytoremediation capacity and behaviours of *Panicum aquaticum* Poir (Figure 29) at different Pb concentrations (0.0, 0.5, 1.0, 2.0, 4.0 and 8.0 mM) have investigated. At highest Pb levels, endoderm thickness increased due to the decrease in shoot deposition and translocation. Increased absorption of Pb (2279 $\mu g\ g^{-1}\ DW^{-1}$) has occurred in roots with increasing Pb concentrations in plant roots. These results proved that this plant is a suitable plant to be used in phytoremediation studies in Pb polluted areas (Pires-lira et al. 2020).

Vetiveria zizanioides (Linn.) Nash, shown in Figure 30, has been evaluated in soil polluted with Cd, Pb, Cu, Zn heavy metals (single and/or mixed of these heavy metals). The plant growth, metal uptake and accumulation by various plant parts have been examined. At plant height, no significant difference ($p > 0.05$) has detected between all single and mixed heavy metal-containing plants in comparison to the control plant. But a considerable extent higher ($p < 0.05$) accumulation has detected in roots, tubers, and the total accumulation of each metal in proportion to other doses. *Vetiveria zizanioides* (Linn.) Nash has grown in soils containing all heavy metals has the highest Zn (3322 ± 21.6 mg kg-1), then Cu (430 ± 11.4 mg kg-1), Pb (197 ± 13.5 mg kg-1), and Cd (100 ± 0.7 mg kg-1), respectively. *Vetiveria zizanioides* (Linn.) Nash has grown in soils containing all heavy metals and accumulated advanced level of heavy metal than soils containing only one heavy metal. Furthermore, the lower roots and substrates of *Vetiveria zizanioides* (Linn.) Nash has been observed to exhibit a powerful tendency for further intake and accumulation of Cd, Pb, Cu and Zn metals in single contaminated and/or mixed contaminated soils (Ng et al. 2020).

Figure 29. *Panicum aquaticum* Poir (Rodrigues, 2013).

Figure 30. *Vetiveria zizanioides* (Linn.) Nash (Gingade, Varghese, and Manivel 2015).

Araceae Family

The *Araceae* family is one of the most diverse monocotyledons, comprising approximately 105 genera and 3,300 species. They have gained economic importance thanks to their ornamental and medicinal use and also edible. The most important therapeutic species of this family, which are used in the treatment of many diseases to be used in the treatment of high fever caused by malaria in the Amazon region, belong to the genus

Philodendron Schott. The plant known as *Colocasia esculenta* (L.) Schott rhizomes belongs to this family and is consumed frequently as a foodstuff and food stock for animals. Some ornamental plants belonging to the *Araceae* family are *Anthurium Schott*, *Philodendron Schott*, *Dieffenbachia Schott*, *Monstera Adams* and *Zantedeschia Spreng* (Frausin et al. 2015). The fruits of *Mostera*, one of The other plants belonging to the *Araceae* family, are used as a food source. The *Colocasia esculenta* tubers are used in the production of starchy baby foods, in the treatment of dyspepsia and diarrhea from Acorus calamus rhizomes, and the steam juice obtained from the body of *Alocasia macrorrhiza* is used to reduce the effect of scorpion bite poison (Mohit 2019).

A study was carried out to investigate the usability of *Pistia stratiotes* L. plant (Figure 31) in removing Fe, C, and Hg heavy metals from the aqueous environment. In this study, it has been applied at 5, 10 and 15 mg L^{-1} concentrations, being the same for each heavy metal. The highest heavy metal accumulation in plants was measured as 53.20% of Cu^{2+}, 83.20% of Fe^{3+} and 62.14% of Hg^{2+}in plants where 5 mg L^{-1} concentration was applied. Also, maximum fresh plant biomass, plant root length, growth rate and chlorophyll content occurred in plants growing with a concentration of 5 mg L^{-1} of pollutants. This study proves that *P. stratiotes* plant is an effective accumulator at 5 mg L^{-1} and has remediation capacity for Cu^{2+}, Fe^{3+}, Hg^{2+} heavy metals. This plant is thought to help clean synthetic and industrial wastewater without interfering with the ecosystem (V. Kumar et al. 2019).

Figure 31. *Pistia stratiotes L.* (Sandilyan et al. 2018); Howcard, 2020).

Figure 32. *Lemna minor* (Armstrongs 2015).

Duckweed that shown in Figure 32, is a plant that grows in aqueous media such as streams, lakes, and other water sources. It is a widely preferred plant in ecotoxic studies small size, simple and easily understandable morphology, short life cycle and susceptibility to environmental contaminants. It is used in the accumulation and removal of heavy metals and in determining the toxicity occasioned by heavy metals in Cd, Co, Ni, Zn, and Cu (Lu et al. 2018).

In another study, the accumulation capacity of the *Lemna minor* plant was tested for boron (B) heavy metal under salt stress. In the experiment, 2 mg L^{-1} B and 0-200 mM sodium chloride (NaCl) were used. The results showed that, depending on the increasing salt concentration, the B accumulation capacity of the plant first decreased and then increased. It was concluded that *L. minor* plant can be used for B accumulation at salt concentrations below 100 mM (C. Liu et al. 2018).

Daud et al. have investigated the usability of duckweed (*L. minor*) plant for the phytoextraction method in removing heavy metals from Zn, Cu, Pb, Fe and Ni from landfill leachate. For this, they have examined the plant by testing it every 3 days for 2 weeks in the presence of heavy metal concentrations. The results of the study have shown that duckweed significantly reduced the heavy metal concentration in the landfill leachate. As a result of the obtained data, duckweed has been identified as a retainable alternative plant for the phytoextraction method and also, the plant has been proposed for the cleanup of wastewater pollutants in sanitary landfill leachate (Daud et al. 2018).

Figure 33. *Azolla filiculoides* (Vanderhoff 2011).

Using two aquatic macrophyte *Lemna minor* and *Azolla filiculoides* plants grown in a semi-arid area of Ethiopia, eight different heavy metal accumulation had been investigated for each plant. Consequently, heavy metal accumulation has observed in both plants. Compared to other heavy metals, the highest BF values have detected in manganese and iron heavy metals at both plants. The *L. minor* plant has accumulated a high amount for cobalt, zinc, manganese, and iron heavy metals and the medium amount for cadmium, copper, nickel, and chrome. The *A. filiculoides* (Figure 33) has shown high accumulator performance for iron, copper, manganese, and zinc heavy metals. The same plant has exhibited moderate deposition performance for cobalt, chrome, and nickel and slight amount deposition performance for cadmium. In the present study, Amare et al. have been observed better relationship between heavy metal accumulations in plant parts and aquatic environment for *A. filiculoides* compared to *L. minor* plant (Amare, Kebede, and Mulat 2018).

Rana and Maiti have been researched the copper, manganese, chrome, zinc, cadmium, cobalt, nickel, and lead heavy metals cleanup potential with *Colocasia esculenta* (L.) Schott plant. In the study, the usability of this plant for municipal wastewater treatment was investigated. While heavy metals of copper, cadmium, cobalt, lead and nickel accumulated in root tissue, heavy metals of manganese and zinc accumulated in shoot tissues. It has been demonstrated that *Colocasia esculenta* (L.) Schott can be used for the stabilization of copper, cadmium, cobalt, nickel and lead heavy metals. In the study, the bioaccumulation factor has calculated higher than 1 and the translocation factor has been calculated less than 1. And also, the researchers concluded that *Colocasia esculenta* (L.) Schott can be used for the improved of municipal wastewater including the Mn and Zn heavy metals when considered to their bioconcentration (BCF > 1) and translocation factor (TF > 1) values (Rana and Maiti 2018).

Abbas et al. The use of the indoor plant *Epipremnum aureum* Engl, a member of the *Araceae* family, in the phytoremediation of Zn heavy metal, which has toxic effects in high concentrations, has been investigated. Different concentrations of $ZnCl_2$ were applied in the study. Doses of 10 ppm, 20 ppm and 30 ppm doses of $ZnCl_2$ caused chlorophyll degradation in the plant and inhibition of vascular growth in the leaf and stem. However, Zn accumulation has occurred in roots and leaves while no apparent Zn toxicity has observed in roots. This situation has shown that the plant can carry Zn to aboveground parts. The *Epipremnum aureum* Engl. (Figure 34) has been reported that the plant can be used in phytoremediation for Zn heavy metal (Abbas, Mehnaz, and Saleem Khan 2019).

Figure 34. *Epipremnum aureum* Engl (Flora Fauna Web 2019).

Euphorbiaceae **Family**

The genus *Euphorbia* (*Euphorbiaceae* family) is used all over the world in medicine, traditional Chinese medicine. Members of these taxa are immediately recognized by their special clusters and latex. The chemical composition of essential oils obtained from *Euphorbia* species is used in the treatment of many diseases, especially inflammation, microbial infections, and cancer (Salehi et al. 2019).

Ricinus communis (castor oil) (Figure 35), which belongs to the *Euphorbiacea*e family, is known as castor oil and it is one of the commercially important plants that used as biofuel. These commonly found plants can survive in areas of stress and environmental pollution. To evaluate the effectiveness of this herb in phytoremediation of heavy metals, an in-vitro pot-based study was conducted. In the study, Pb accumulation with different plant parts of experimental and control castor oil plants was evaluated. The collected plant samples were analyzed for heavy metal deposition by Atomic Absorption Spectroscopy (AAS). Lead showed the greatest accumulation in the vegetative parts of the plant above ground. Lead accumulation was found as branch > root > leaf in the sample facility grown in fly ash modified soil. However, it was determined that lead accumulation in garden soil was quite low and the order of accumulation amount was leaf > root > shoot. It was concluded that Ricinus communis plant has phytoremediation potential in contaminated areas (Roychowdhury et al. 2019).

It is known that the application of *Ricinus communis* for plant breeding instead of edible and non-edible stress-sensitive plants can be a useful choice for improving polluted soil. And also, *R. communis* uses as medicinal products, carbon sequestration, reduction of greenhouse gases (GHG), increasing soil fertility and erosion. *R. communis*, it has been found that it has a perfect talent to remove most of the toxic metals as Cd, Pb, Ni, As, Cu and also pesticides as well as some organic pollutants from the soil (Bauddh et al. 2015).

Figure 35. *Ricinus communis* (Marwat and Fazal-Ur-rehman 2017).

In a study conducted in Cuba, the hyperaccumulation capacity of some plant samples known as nickel hyperaccumulators has been evaluated that has collected from their natural ultramafic environment. Different plant samples such as *Phyllanthus orbicularis* Kunth and *Euphorbia helenae* Urb. s. l. (Figure 36), *Phyllanthus discolour* Poepp. ex Spreng, *Phyllanthus pallidus* Müll. Arg., *Leucocroton flavicans* Müll. Arg., and *Leucocroton lineralifolius* Britt have been examined in the study. The *Leucocroton flavicans* (6710-15500 mg kg^{-1}) and *Leucocroton linearifolius* (13310-27240 mg kg^{-1}) plants have accumulated Ni heavy metal in their leaves. *Euphorbia helenae* s., *Phyllanthus orbicularis*, *Phyllanthus discolour* and *Phyllanthus pallidus* plants have accumulated 2000-9000, 4140-10950, 13670-31490 and 15390-60170 mg kg^{-1} in their leaf's Ni heavy metal, respectively. The same plants have accumulated 30900, 3660, 14940 and 17150 mg kg^{-1} Ni on their branches (Berazaín et al. 2007).

Jatropha curcas (Figure 37) is connected to the *Euphorbiaceae* family and it is a perennial medicinal plant widely used in plant breeding practices and biodiesel production. In this study, a Pb (lead) has applied to *J. curcas* seedlings with different concentration. One week after the Pb application, Pb content in the plants has measured by atomic absorption spectrophotometer. It has observed that Pb accumulation in the plant increased significantly with the rise of metal in the growth medium. The highest Pb accumulation has reported being the plant with 400 mg kg^{-1} dry weight that in plants treated with 40mg L^{-1} (Atito, Heneidak, and El-sayed 2018).

Figure 36. *Phyllanthus orbicularis* Kunth, *Euphorbia helenae* Urb. s. l., respectively (Leyva, Vales, & Francisco, 2018; Steinmann, Van Ee, Berry, & Gutiérrez, 2007).

Madruga-Negrete et al. have conducted a study to examine the phytoremediation behaviour of the *Jatropha curcas* plant in mercury-contaminated soils. In the experiment, different mercury concentrations have used as 0.1.5 and 10 μg g^{-1} Hg. Changes in mercury content in plant roots, leaves and stems have observed for 4 months. At the end of 4 months, growth behaviour, translocation factor and bioconcentration factor values of the plant and mercury accumulation in the tissues have examined. Mercury accumulation in tissues has determined as roots> leaves> stems, respectively. The better metal accumulation has occurred among the 2nd and 3rd months of incubation. The highest has been calculated in 2nd month to between 0.79 and 1.04 mercury concentrations. On the other hand, the calculated between 0.21 and 1.43 mercury concentrations. This translocation factor and bioconcentration factor values that obtained with *J. curcas* studies prove that they can be used to improve mercury-contaminated soils (Marrugo-Negrete et al. 2015).

Figure 37. *Jatropha curcas* (iNaturalist 2020).

Violaceae Family

The *Violaceae* family, which consists of about 500 species extensively distributed around the world, has been used in Chinese medicine for various treatment, including blood purification and caries and ulcer treatment. *Violaceae* family, which has important medicinal plants partaking in traditional medicine, is also used for diseases such as cough, fever, flu malaria, cold and is also used as an anticancer drug (Chandra et al. 2015).

Wu et al. have determined that there is not enough information about *Viola baoshanensis*, which is known as the Cd hyperaccumulator, to the accumulation of lead or zinc. For that reason, they have been conducted studies to identify whether *V. baoshanensis* could accumulate lead or zinc through pot and hydroponics experiments. Moreover, more detailed information about the known Cd hyperaccumulation properties of these strain has been researched. When the plants were examined at the end of the study, it was determined that approximately 1090 $mgkg^{-1}$ Cadmium, 1902 $m\ kg^{-1}$ lead and 3428 $mgkg^{-1}$ Zinc have accumulated in the shoots of *V. baoshanensis* plant. In experiments, *V. baoshanensis* has shown high tolerance to whole heavy metals, and also the results of three elements

(>2% Cd on dry matter basis, >1% Pb and >0.5% Zn) were obtained in their shoots. These results have shown that *V. baoshanensis* is a powerful Pb and Zn accumulator as well as being a Cadmium hyperaccumulator (Wu et al. 2010).

The behaviour of some *Viola* endemic species has been studied in an abandoned mine land containing mainly As in Alchar, Republic of Macedonia. It has been observed that the amount of As in the roots of *Viola macedonica* varies between 78- 2124 mg kg^{-1} arsenic. *Viola allchariensis* accumulated less than 100 mg kg^{-1} of As in aboveground parts and also, 187.439 mg kg^{-1} arsenic accumulated in its shoots. The accumulation of arsenic in the roots of *Viola* species is making these plants valuable for bioindication and phytostabilization (plant stabilization) in naturally charged and anthropogenically polluted soils with arsenic (Stevanović et al. 2010).

Figure 39. *Hybanthus enneaspermus* (Rajsekhar et al. 2016), (Wahlert et al. 2014).

The extraordinary biodiversity, soil types, and climatic value ranges found in Papua New Guinea suggest that there is a high probability of high metal hyper-accumulating plants in this area. To prove this, systematic quantitative evaluation of hyperaccumulation with X-ray fluorescent devices by taking a large number of herbarium samples was performed in a non-destructive and effective manner. At the end of the study, it was determined that *Rinorea sp* plant has hyperaccumulation capacity. It was observed that the plant accumulated Zn, Ni and Mn heavy metals as 104,

171 and 459 μg g^{-1}, respectively. These results show that *Rinorea sp.* can be used as a hyperaccumulator to remove metal contents found in contaminated soils (Do et al. 2020). In previous studies, when the elements accumulated in the leaves of hyperaccumulator plants and the elements in the soil chemistry were compared, it was observed that there was a significant correlation between the variable Ca, K, and P in the soil and the Ca, K and P in the leaves. Determining the hyperaccumulators of Ni is necessary to protect plants from the toxic effects of Ni heavy metal and facilitate their potential use in the future. In this study, it has been reported that Ni metal can accumulate in the underground parts of *Rinorea bengalensis* (12,800 μg g^{-1}), *Rinorea javanica* (9680 μg g^{-1}) and *Rinorea sp* (5830 μg g^{-1}). It has been determined that these plants are suitable for the cleaning of polluted soil with Ni (van der Ent, Erskine, and Sumail 2015). At least seven taxa from the genus *Hybanthus* are known to as hyper-accumulative plants for Ni heavy metal. A study has conducted to evaluate the Ni accumulation capacity of *Hybanthus enneaspermus* plant. Ni content has determined as 29 and 1786 μg g^{-1} dry weight in soil and plant, respectively. Study results have shown that *Hybanthus enneaspermus* plant can be used for phytoremediation of polluted soils with Ni (Rajakaruna and Baker 2004).

Polygonaceae Family

Polygonaceae family known as smartweed or knot wood family is a family of flowering plants found all over the world. The genera with the most members are *Eriogonum*, *Persicaria*, *Cocoloba* and *Rumex*. It has been reported that herbs belonging to this family are widely used in the treatment of many medical conditions such as asthma, cough, diarrhoea, eczema, ear pain, jaundice, kidney disease, toothache, leprosy, intestinal parasites, and inflammatory disease. *Rumex*, one of the most important genera of the *Polygonaceae* family, is widely available in the world. About 200 species of the *Rumex* genus are used in medicine. Parts of the plants such as root, leaf, seed, aboveground parts and plant water are used. Also, they contain various pharmacological activities such as antioxidant, cytotoxic, antifertility, laxative, antifungal, antipyretic, antiviral, antifungal, antidiarrhoeal and antibacterial (Tonny, Sultana, and Siddika 2017).

The relationship between Cr, Cd, Pb and Mn in dock (*Rumex patientia* L.) plant was investigated using plant breeding method in polluted soils with heavy metals. In the DTPA soil test, dock (*Rumex patientia* L.) plant was investigated in terms of the effects of Mn interactions with EDTA chelate application using some heavy metals. Approximately 100 mg kg^{-1} Cr (Cr $(NO_3)_3$), Cd (Cd $(NO_3)_3$) and Pb (Pb $(NO_3)_2$) were applied as heavy metals. At the beginning of the flowering of the dock plant, 0, 5, 10 and 15 mmol kg^{-1}doses of EDTA were applied to the pots. Metal contents in different concentrations were measured in roots and shoots of the harvested plant after two months of the growth period. It was observed to *Rumex patient* L. plant (Figure 40) is suitable to remove the heavy metal pollution with phytoremediation (Adiloğlu 2017).

Figure 40. Duck (*Rumex patientia* L.) plant (Orijinal).

It has been reported in previous studies that cadmium can be removed from agricultural soil using *Rumex patientia* L. In pot experiments conducted under greenhouse conditions at Namık Kemal University campus, 100 $mgkg^{-1}$ cadmium and different applied of EDTA were applied to the soil. Plants have been harvested after two months of incubation. In the harvested plants, it has observed that with the increase of the applied EDTA dose, the yield in plant roots and shoots decreased, but the Cd content increased 1% significance level. The results have been shown that *Rumex patientia* L. plant can be grown in the phytoremediation of the Cd heavy metal in soil (Adiloğlu, Adiloğlu, et al. 2016).

CONCLUSION

Increasing industrial activities in recent years, pesticide use, rapidly increasing population and accompanying urbanization increase heavy metal pollution in the ecology. Increasing heavy metal pollution affects living beings directly or indirectly due to its toxic effects and accumulation in their environment. In order to prevent or minimize this pollution, studies have increased in recent years. Phytoremediation has known as the most preferred method for naturally removing heavy metals and some other pollutions from the environment with the help of plants. The use of hyperaccumulator plants in the removal of heavy metals is the most used method in this regard, and hundreds of hyperaccumulator plants belonging to many plant families have been identified so far.

Plants have a key role in maintaining the balance in the ecosystem thanks to the metabolic events they perform and the interactions they establish with microorganisms. Hyperaccumulator plants are successful in removing heavy metals from the soil and water resources in various ways and reusing these resources. However, increasing pollutant concentrations are affecting the plant's success in phytoremediation, limiting the use of technology in highly polluted areas. In addition, the resistance of the plant against toxicity is decreased with increasing concentrations and the accumulation capacity is decreased with the plant system deterioration.

Although the use of hyperaccumulator herbs is effective in phytoremediation, it is not sufficient alone. Therefore, it is very important to investigate and define mechanisms such as heavy metal translocation, uptake, and detoxification in hyperaccumulator plants and to improve the accumulation capacity of these plants. For example, microorganisms in the soil are provided rapid growth of the plant by increasing the usefulness of heavy metals in the soil, improving the soil quality, and are helping the plant to adapt easily. In order to increase the applicability and effect of plant breeding in preventing environmental pollution, efforts should be focused on improving the performance of species used for this purpose. Within this, it is extremely important to research and conduct different applications such as genetic studies and bacteria supported-approach in order to realize more effective phytoremediation applications in the future.

REFERENCES

Abbas, Razi, Samina Mehnaz, and Aisha Saleem Khan. 2019. "Evaluation of Potential of Epipremnum Aureum Engl. In Removing Zinc (Zn) Toxicity." *Pakistan Journal of Botany* 51 (5): 1599–1604.

Abdel-Naime, Waleed A, John Refaat Fahim, Mohamed Salah Kamel, and Mostafa A Fouad. 2016. "Botanical Studies of the Leaf of *Melissa Officinalis* L., Family: Labiatae, Cultivated in Egypt." *Journal of Pharmacognosy and Phytochemistry JPP* 5 (56): 98–104.

Abid, Rafia, Seema Mahmood, Shazia Ghaffar, Saman Zahra, and Sibgha Noreen. 2019. "Tannery Effluent Induced Morpho-Biochemical Expressions and Chromium Accumulation in *Jatropha Curcas* L. And Pongamia Pinnata L." *Sains Malaysiana* 48 (5): 927–36.

Adiloğlu, Sevinç. 2016. Using Phytoremediation with Canola to Remove Cobalt from Agricultural Soils. *Polish Journal of Environmental Studies* 25 (6): 2251–54. https://doi.org/10.15244/pjoes/62138.

Adiloğlu, Sevinç. 2017. The Effects of Manganese on the Remediation of the Heavy Metal Contaminated Soil by Using the Dock (*Rumex Patientia* L.) Plant." *Desalination and Water Treatment* 93 (3): 335–38.

Adiloğlu, Sevinç, Aydın Adiloğlu, Funda Eryilmaz Açıkgöz, Tuba Yeniaras, and Yusuf Solmaz. 2016. Phytoremediation of Cadmium from Soil Using Patience Dock (*Rumex Patientia* L *Analytical Letters* 49 (4): 601–6.

Adiloglu, Sevinç, and Mehmet Turgut Saglam Ali Süme. 2015. "Chrome (Cr) Pollution in Agricultural Areas Improvement by Phytoremediation Method with Canola (*Brassica Napus* L.) Plant Growing." *Journal of Essential Oil-Bearing Plants* 18 (5): 1180–86.

Adiloğlu, Sevinç, M. Turgut Sağlam, Aydın Adiloğlu, and Ali Süme. 2016. "Phytoremediation of Nickel (Ni) from Agricultural Soils Using Canola (Brassica Napus L.)." *Desalination and Water Treatment* 57 (6): 2383–88.

Adiloğlu, Sevinç, and Göker, Merve. 2021. Phytoremediation: Elimination of Hexavalent Chromium Heavy Metal Using Corn (*Zea Mays* L.). *Cereal Research Communications*. doi: 10.1007/s42976-020-00070-9.

Adiloğlu, Sevinç. 2021. Relation of Chelated Iron (EDDHA-Fe) Applications with Iron Accumulation and Some Plant Nutrient Elements in Basil (*Ocimum Basilicum* L.). *Polish Journal of Environmental Studies*. 30(3).

Akulova, Zoya. 2018. *Brassica Oleracea Var. Acephala*. 2018.

Amare, Elfu, Fassil Kebede, and Worku Mulat. 2018. "Wastewater Treatment by Lemna Minor and Azolla Filiculoides in Tropical Semi-Arid Regions of Ethiopia." *Ecological Engineering* 120 (May): 464–73. https://doi.org/10.1016/j.ecoleng.2018.07.005.

Angelova, V. R., R. I. Ivanova, J. M. Todorov, and K. I. Ivanov. 2017. "Potential of Rapeseed (*Brassica Napus* L.) for Phytoremediation of Soils Contaminated with Heavy Metals." *Journal of Environmental Protection and Ecology* 18 (2): 468–78.

Angelova, Violina. 2013. "Potential of Some Medicinal and Aromatic Plants for Phytoremedation of Contaminated with Heavy Metals Soils." In *BMPC XV*, 1045–48.

Angelova, Violina R., Dimitar F. Grekov, Veselin K. Kisyov, and Krasimir I. Ivanov. 2015. "Potential of Lavender (*Lavandula Vera* L.) for Phytoremediation of Soils Contaminated with Heavy Metals." *Phytoremediation: Management of Environmental Contaminants,*

Volume 1 9 (5): 133–55. https://doi.org/10.1007/978-3-319-10395-2_10.

Aram, H., H. Negahdarı, and L. Tajgardoon. 2019. "Evaluation of Phytoremediation Potential of Rosemary Plant in Contaminated Soils with Cadmium and Zinc." *Journal of Global Ecology and Environment* 9 (2): 54–59. http://www.ikprress.org/index.php/JOGEE/article/view/4784.

Armstrongs, Wayne. 2015. "*Plant Biodiversity Conservatory and Research Center.*" Uconn-University of Connecticut. 2015. http://florawww.eeb.uconn.edu/198502634.html.

Atito, Eman, Samia Heneidak, and Magdi A El-sayed. 2018. "Effect of Lead (Pb) on Phytochemical Variability of *Jatropha Curcas* (L.): A Versatile Perennial of Euphorbiaceae Family." *Journal of Biological Studies* 1 (2): 90–97.

Australian National Botanic Gardens. 2016. *Botanical Web Portal.* An Australian Government Initiative. 2016. http://www.anbg.gov.au/jmiller/factsheets/Vachellia/campechiana_f._campechiana.htm.

Bakali, Abdelmonaim Homrani. 2016. *Plant Biodiversity of South-Western Morocco.* 2016. https://www.teline.fr/en/photos/asteraceae /atractylis-serratuloides#photo-10.

Barceló, Joan, and Charlotte Poschenrieder. 2003. "Phytoremediation: Principles and Perspectives." *Contributions to Science* 2 (3): 333–44. https://doi.org/10.2436/cs.v0i0.310.

Barouchas, Pantelis E, Anastasia Akoumianaki-ioannidou, Aglaia Liopa-tsakalidi, and Nicholas K Moustakas. 2019. *Nickel (Ni) Effects on Shoots and Roots Dry Matter and on Ni Concentration in Shoots of Mojito (Mentha × Villosa) and Lavender (Lavandula Angustifolia)*, 3–5. https://doi.org/10.15835/buasvmcn-hort.

Bauddh, Kuldeep, Kripal Singh, Bhaskar Singh, and Rana P. Singh. 2015. "Ricinus Communis: A Robust Plant for Bio-Energy and Phytoremediation of Toxic Metals from Contaminated Soil." *Ecological Engineering* 84: 640–52. https://doi.org/10.1016/j.ecoleng.2015.09.038.

Bayouli, Ines Terwayet, Beatriz Gómez-Gómez, Houssem Terwayet Bayouli, Teresa Pérez-Corona, Erik Meers, Emna Ammar, Ali Ferchichi, and Yolanda Madrid Albarrán. 2020. "Heavy Metal Transport and Fate in Soil-Plant System: Study Case of Industrial Cement Vicinity, Tunisia." *Arabian Journal of Geosciences* 13 (2). https://doi.org/10.1007/s12517-019-4898-7.

Berazaín, Rosalina, Vicenta De La Fuente, Lourdes Rufo, Nuria Rodríguez, Ricardo Amils, Blanca Díez-Garretas, Daniel Sánchez-Mata, and Alfredo Asensi. 2007. "Nickel Localization in Tissues of Different Hyperaccumulator Species of Euphorbiaceae from Ultramafic Areas of Cuba." *Plant and Soil* 293 (1–2): 99–106. https://doi.org/10.1007/s11104-007-9227-3.

Bianucci, Eliana, Juan Manuel Peralta, Ana Furlan, Luis E. Hernández, and Stella Castro. 2020. "Arsenic in Wheat, Maize, and Other Crops." *Arsenic in Drinking Water and Food*, 279–306. https://doi.org/10.1007/978-981-13-8587-2_9.

BitkiVT Online Bitki Veritabanı-İTÜ. 2020. "BitkiVT - Lavandula Angustifolia Mill. Detayları." İstanbul Teknik Üniversitesi Peyzaj Mimarlığı. 2020. http://www.bitkivt.itu.edu.tr/vt/report.php?sor=174.

Carrubba, Alessandra, Caterina Catalano, and Italian National. 2009. *Climate Change, Intercropping, Pest Control and Beneficial Microorganisms. Climate Change, Intercropping, Pest Control and Beneficial Microorganisms*. https://doi.org/10.1007/978-90-481-2716-0.

Chandra, Deepak, Gunjan Kohli, Kundan Prasad, G. Bisht, Vinay Deep Punetha, K.S. Khetwal, Manoj Kumar Devrani, and H. K. Pandey. 2015. "Phytochemical and Ethnomedicinal Uses of Family Violaceae." *Current Research in Chemistry* 7 (2): 44–52. https://doi.org/10.3923/crc.2015.44.52.

Chen, Guohu, Fanli Zeng, Jian Wang, Xinyu Ye, Shidong Zhu, Lingyun Yuan, Jinfeng Hou, and Chenggang Wang. 2019. "Transgenic Wucai (*Brassica Campestris* L.) Produced via Agrobacterium-Mediated Anther Transformation in Planta." *Plant Cell Reports* 38 (5): 577–86. https://doi.org/10.1007/s00299-019-02387-0.

Daud, M. K., Shafaqat Ali, Zohaib Abbas, Ihsan Elahi Zaheer, Muhammad Ahsan Riaz, Afifa Malik, Afzal Hussain, Muhammad Rizwan, Muhammad Zia-Ur-Rehman, and Shui Jin Zhu. 2018. "Potential of Duckweed (Lemna Minor) for the Phytoremediation of Landfill Leachate." *Journal of Chemistry* 2018. https://doi.org/10.1155/2018/3951540.

Do, Christina, Farida Abubakari, Amelia Corzo Remigio, Gillian K. Brown, Lachlan W. Casey, Valérie Burtet-Sarramegna, Vidiro Gei, Peter D. Erskine, and Antony van der Ent. 2020. "A Preliminary Survey of Nickel, Manganese and Zinc (Hyper)Accumulation in the Flora of Papua New Guinea from Herbarium X-Ray Fluorescence Scanning." *Chemoecology* 30 (1). https://doi.org/10.1007/s00049-019-00293-1.

Dökmeci, A. H., Adiloğlu S. 2020. The Phytoremediation of Chrome From Soil Using (*Cirsium Vulgare*) and the Health Effects of Bioaccumulation. *Biosciences Biotechnology Research Asia*, 17(3): 535-541.

Drobot, Paul. n.d. *Symphytum Officinale L.* Wisconsin State Herbarium, UW-Madison Department of Botany. http://wisflora.herbarium.wisc.edu/imagelib/imgdetails.php?imgid=18469.

Echeverría, María L., Sara I. Alonso, and Viviana M. Comparatore. 2017. "Survey of the Vascular Plants of Sierra Chica, the Untouched Area of the Paititi Natural Reserve (Southeastern Tandilia Mountain Range, Buenos Aires Province, Argentina)." *Check List* 13 (6): 1003–36. https://doi.org/10.15560/13.6.1003.

Ent, Antony van der, Peter Erskine, and Sukaibin Sumail. 2015. "Ecology of Nickel Hyperaccumulator Plants from Ultramafic Soils in Sabah (Malaysia)." *Chemoecology* 25 (5): 243–59. https://doi.org/10.1007/s00049-015-0192-7.

EPA, US. 1999. *Phytoremediation Resource Guide*. Washington, DC. https://www.epa.gov/sites/production/files/2015-04/documents/phytoresgude.pdf.

Etim, E E. 2012. "Phytoremediation and Its Mechanisms: A Review." *International Journal of Environment and Bioenergy International*

Journal of Environment and Bioenergy Journal Homepage: Www.ModernScientificPress.Com 2 (3): 120–36.

Flora Fauna Web. 2019. *Epipremnum Pinnatum (L.) Engl.* A Singapore Government Agency Website. 2019. https://www.nparks.gov.sg/florafaunaweb/flora/1/3/1393.

Flora of Eastern Washington and Adjacent Idaho. 2014. *Puccinellia Distans.* Eastern Washington University. 2014. https://inside.ewu.edu/ewflora/category/puccinellia/.

Frausin, G., R. B. S. Liam, A. F. Hidalgo, L. C. Ming, and A. M. Pohlit. 2015. "Plants of the Araceae Family for Malaria and Related Diseases: A Review." *Revista Brasileira de Plantas Medicinais* 17 (4): 657–66. https://doi.org/10.1590/1983-084x/14_024.

Gallaher, Timothy, and Mark Merlin. 2010. " Biology and Impacts of Pacific Island Invasive Species. 6. *Prosopis Pallida* and *Prosopis Juliflora* (Algarroba, Mesquite, Kiawe) (Fabaceae) ." *Pacific Science* 64 (4): 489–526. https://doi.org/10.2984/64.4.489.

Ganesan, Kumar, and Baojun Xu. 2018. "A Critical Review on Phytochemical Profile and Health Promoting Effects of Mung Bean (Vigna Radiata)." *Food Science and Human Wellness* 7 (1): 11–33. https://doi.org/10.1016/j.fshw.2017.11.002.

Gebru, Mehari Girmay. 2019. "Phytoremediation of Heavy Metals Released from Mining Waste Drainage Using Selected Plant Species, in Ethiopia." *Budapest International Research in Exact Sciences (BirEx) Journal* 1 (3): 1–4. https://doi.org/10.33258/birex.v1i3.343.

Gholinejad, Bahram, Shima Khashij, Farshid Ghorbani, Isa Bandak, and Asghar Farajollahi. 2020. "Effects of Lead Ions on Germination, Initial Growth, and Physiological Characteristics of *Lolium Perenne* L. Species and Its Bioaccumulation Potential." *Environmental Science and Pollution Research*. https://doi.org/10.1007/s11356-019-06766-8.

Gingade, Smitha, Thania Sara Varghese, and Ponnuchamy Manivel. 2015. *Cultivation of Vetiver [Vetiveria Zizanioides (Linn)],* no. June 2016: 1–16.

Gómez-Arroyo, Sandra, Josefina Cortés-Eslava, Paola Loza-Gómez, Francisco Arenas-Huertero, Michel Grutter de la Mora, and Ofelia

Morton Bermea. 2018. "In Situ Biomonitoring of Air Quality in Rural and Urban Environments of Mexico Valley through Genotoxicity Evaluated in Wild Plants." *Atmospheric Pollution Research* 9 (1): 119–25. https://doi.org/10.1016/j.apr.2017.06.009.

Grignet, Arnaud, Annette de Vaufleury, Arnaud Papin, and Valérie Bert. 2019. "Urban Soil Phytomanagement for Zn and Cd in Situ Removal, Greening, and Zn-Rich Biomass Production Taking Care of Snail Exposure." *Environmental Science and Pollution Research*. https://doi.org/10.1007/s11356-019-06796-2.

Harvard Health Publishing. 2018. *Legumes: A Quick and Easy Switch to Improve Your Diet*. Harvard Medical School. 2018. https://www.health.harvard.edu/staying-healthy/legumes-a-quick-and-easy-switch-to-improve-your-diet.

Hesami, Reza, Azam Salimi, and Seyed Majid Ghaderian. 2018. "Lead, Zinc, and Cadmium Uptake, Accumulation, and Phytoremediation by Plants Growing around Tang-e Douzan Lead–Zinc Mine, Iran." *Environmental Science and Pollution Research* 25 (9): 8701–14. https://doi.org/10.1007/s11356-017-1156-y.

Howard, Richard A. 2020. *Pistia Stratiotes L. - Water Lettuce*. Smithsonian Institution. 2020. Russian Federation, Saint Petersburg, Komarov.

Ibañez, Sabrina, María I. Medina, and Elizabeth Agostini. 2020. "Vicia : A Green Bridge to Clean up Polluted Environments." *Applied Microbiology and Biotechnology*, 13–21. https://doi.org/10.1007/s00253-019-10222-5.

iNaturalist. 2020. *Jatropha Curcas*. 2020. https://www.inaturalist.org/observations/40171986.

Keeran, Nisha Surendran, Usha Balasundaram, Ganesan Govindan, and Ajay Kumar Parida. 2019. *Prosopis Juliflora. Transgenic Plant Technology for Remediation of Toxic Metals and Metalloids*. Elsevier Inc. https://doi.org/10.1016/b978-0-12-814389-6.00018-3.

Khan, Raees, Sheikh Zain, Ul Abidin, Abdul Samad Mumtaz, Shayan Jamsheed, and Hadayat Ullah. 2017. *Comparative Leaf and Pollen*

Micromorphology on Some Grasses Taxa (Poaceae) Distributed in Pakistan 1 (December): 72–82.

Kök, Aysu Başak, Mehmet Direnç Mungan, Sami Doğanlar, and Anne Frary. 2020. "Transcriptomic Analysis of Selenium Accumulation in Puccinellia Distans (Jacq.) Parl., a Boron Hyperaccumulator." *Chemosphere* 245. https://doi.org/10.1016/j.chemosphere.2019.125665.

Kumar, Dharmendra, Durgesh Kumar Tripathi, Shiliang Liu, Vivek Kumar Singh, Shivesh Sharma, Nawal Kishore Dubey, Sheo Mohan Prasad, and Devendra Kumar Chauhan. 2017. "*Pongamia Pinnata* (L.) Pierre Tree Seedlings Offer a Model Species for Arsenic Phytoremediation." *Plant Gene* 11: 238–46. https://doi.org/10.1016/j.plgene.2017.06.002.

Kumar, Vinod, Jogendra Singh, Ashu Saini, and Pankaj Kumar. 2019. "Phytoremediation of Copper, Iron and Mercury from Aqueous Solution by Water Lettuce (*Pistia Stratiotes* L .)." *Environmental Sustainability*, no. 0123456789. https://doi.org/10.1007/s42398-019-00050-8.

Lajayer, Hemayat Asgari, Gholamreza Savaghebi, Javad Hadian, Mehrnaz Hatami, and Maryam Pezhmanmehr. 2017. "Comparison of Copper and Zinc Effects on Growth, Micro- and Macronutrients Status and Essential Oil Constituents in Pennyroyal (*Mentha Pulegium* L.)." *Revista Brasileira de Botanica* 40 (2): 379–88. https://doi.org/10.1007/s40415-016-0353-0.

Leyva, Luis Manuel, Dennys De Vales, and Gabriela Rijo De Francisco. 2018. *Characterization of the Population Structure of Phyllanthus Orbicularis (Phyllanthaceae) in Sierra de Cajálbana, La Palma, Pinar Del Río, Cuba.*, no. December.

Li, Qing X., and Chiou L. Chang. 2016. *Basil (Ocimum Basilicum L.) Oils. Essential Oils in Food Preservation, Flavor and Safety*. Elsevier Inc. https://doi.org/10.1016/B978-0-12-416641-7.00025-0.

Liu, Chunguang, Wancong Gu, Zheng Dai, Jia Li, Hongru Jiang, and Qian Zhang. 2018. "*Boron Accumulation by Lemna Minor L. under Salt Stress*," 1–6. https://doi.org/10.1038/s41598-018-27343-y.

Liu, Zhouli, Wei Chen, and Xingyuan He. 2018. "Evaluation of Hyperaccumulation Potentials to Cadmium (Cd) in Six Ornamental Species (Compositae)." *International Journal of Phytoremediation* 20 (14): 1464–69. https://doi.org/10.1080/15226514.2018.1501343.

Lu, Qianqian, Tingting Zhang, Wei Zhang, Chunlei Su, Yaru Yang, Dan Hu, and Qinsong Xu. 2018. "Alleviation of Cadmium Toxicity in Lemna Minor by Exogenous Salicylic Acid." *Ecotoxicology and Environmental Safety* 147 (March 2017): 500–508. https://doi.org/10.1016/j.ecoenv.2017.09.015.

Marrugo-Negrete, José, José Durango-Hernández, José Pinedo-Hernández, Jesús Olivero-Verbel, and Sergi Díez. 2015. "Phytoremediation of Mercury-Contaminated Soils by *Jatropha Curcas*." *Chemosphere* 127: 58–63. https://doi.org/10.1016/j.chemosphere.2014.12.073.

Marwat, Sarfaraz Khan, and Fazal-Ur-rehman. 2017. "*Ricinus Cmmunis*: Ethnomedicinal Uses and Pharmacological Activities." *Pakistan Journal of Pharmaceutical Sciences* 30 (5): 1815–27.

Matson, Steve. 2011. *Puccinellia Distans*. University of California, Berkeley. 2011. https://calphotos.berkeley.edu/cgi/img_query?enlarge=0000+0000+0411+0464.

Mishra, Tripti, and Vimal Chandra Pandey. 2019. *Phytoremediation of Red Mud Deposits Through Natural Succession. Phytomanagement of Polluted Sites*. Elsevier Inc. https://doi.org/10.1016/b978-0-12-813912-7.00016-8.

Mohit, Chaturvedi. 2019. "A Review on Plant Family: ARACEAE." *International Journal for Research in Applied Science and Engineering Technology* 7 (2): 1049–56. https://doi.org/10.22214/ijraset.2019.2166.

Morse, Keir. 2016. *Mentha Pulegium L.* University of California, Berkeley. 2016. https://ucjeps.berkeley.edu/eflora/eflora_display.php?tid=33206.

Nadda, Gireesh, Dhananjay Kumar Tewary, Adarsh Shanker, and Virendra Singh. 2012. "*Salvia Sclarea* L. (Lamiales: Lamiaceae) - A New Host Record for *Helicoverpa Armigera* (Hübner) (Lepidoptera: Noctuidae)." *Munis Entomology & Zoology* 7 (1): 642–45. http://www.

munisentzool.org/yayin/vol7/issue1/642-645.pdf%5Cnhttp://www.munisentzool.org/?page=abstract&jid=14&id=550.

NC State University. 2020. *NC State Extension*. 2020. https://plants.ces.ncsu.edu/plants/brassica-juncea/.

Ng, Chuck Chuan, Amru Nasrulhaq Boyce, Mhd Radzi Abas, Noor Zalina Mahmood, and Fengxiang Han. 2020. "Evaluation of Vetiver Grass Uptake Efficiency in Single and Mixed Heavy Metal Contaminated Soil." *Environmental Processes*. https://doi.org/10.1007/s40710-019-00418-2.

Nikolić, Magdalena, and Svetlana Stevović. 2015. "Family Asteraceae as a Sustainable Planning Tool in Phytoremediation and Its Relevance in Urban Areas." *Urban Forestry and Urban Greening* 14 (4): 782–89. https://doi.org/10.1016/j.ufug.2015.08.002.

Nofel, Media. 2016. "Leguminosae Family Tree." *ResearchGate*, no. February: 11–29.

Pandey, Vimal Chandra, Apurva Rai, and John Korstad. 2019. *Aromatic Crops in Phytoremediation: From Contaminated to Waste Dumpsites. Phytomanagement of Polluted Sites*. Elsevier Inc. https://doi.org/10.1016/b978-0-12-813912-7.00009-0.

Pires-lira, Marinês F, Evaristo M De Castro, Jean Marcel, S Lira, Cynthia De Oliveira, Fabrico J Pereira, and Marcio P Pereira. 2020. "Ecotoxicology and Environmental Safety Potential of *Panicum Aquanticum* Poir . (Poaceae) for the Phytoremediation of Aquatic Environments Contaminated by Lead." *Ecotoxicology and Environmental Safety* 193 (February): 110336. https://doi.org/10.1016/j.ecoenv.2020.110336.

Plant Biodiversity Conservatory and Research Center. 2020. *Rosmarinus Officinalis L.* UCONN-UNIVERSITY OF CONNECTICUT. 2020. http://florawww.eeb.uconn.edu/198501280.html.

Rai, Nidhi, Nirmal Kumar Sharma, and Anil Panchal. 2019. "Heavy Metal Accumulation by Selected Plant Species along the National Highway: A Case Study of Udaipur, Rajasthan, India." *International Journal of Environmental Analytical Chemistry* 99 (11): 1078–89. https://doi.org/10.1080/03067319.2019.1616703.

Rajakaruna, N., and A. Baker. 2004. "Serpentine. a Model Habitat for Botanical Research in Sri Lanka." *Ceylon Journal of Science* 32: 1.

Rajsekhar, P. B., R. S. Arvind Bharani, K. Jini Angel, M. Ramachandran, and S. P. V. Rajsekhar. 2016. "*Hybanthus Enneaspermus* (L) F. Muell: A Phytopharmacological Review on Herbal Medicine." *Journal of Chemical and Pharmaceutical Research* 8 (1).

Rana, Vivek, and Subodh Kumar Maiti. 2018. "*Municipal Wastewater Treatment Potential and Metal Accumulation Strategies of Colocasia Esculenta (L.) Schott and Typha Latifolia L. in a Constructed Wetland.*"

Roychowdhury, Roopali, Madhumita Roy, Sufia Zaman, and Abhijit Mitra. 2019. "*Phytoremediation Potential f Castor Oil Plant (Ricinus Communis) Grown on Fly Ash Amended Soil Towards Lead*," no. February.

Sá, Raquel A., Renata A. Sá, Odair Alberton, Zilda C. Gazim, Antonio Laverde, Josiane Caetano, Ana C. Amorin, and Douglas C. Dragunski. 2015. "Phytoaccumulation and Effect of Lead on Yield and Chemical Composition of Mentha Crispa Essential Oil." *Desalination and Water Treatment* 53 (11): 3007–17. https://doi.org/10.1080/19443994.2013.874716.

Salas-Moreno, Manuel, and José Marrugo-Negrete. 2020. "Phytoremediation Potential of Cd and Pb-Contaminated Soils by *Paspalum Fasciculatum* Willd. Ex Flüggé." *International Journal of Phytoremediation* 22 (1): 87–97. https://doi.org/10.1080/15226514.2019.1644291.

Salehi, Bahare, Marcello Iriti, Sara Vitalini, and Hubert Antolak. 2019. "*Euphorbia -Derived Natural Products with Potential for Use in Health Maintenance*," 1–22.

Sandilyan, S, B Meenakumari, A Biju Kumar, and Rupam Mandal. 2018. "Phytoremediation of Copper, Iron and Mercury from Aqueous Solution by Water Lettuce (*Pistia Stratiotes* L.)." *Environmental Sustainability.*

Santoyo-Martínez, Miguel, Patricia Mussali-Galante, Isela Hernández-Plata, Leticia Valencia-Cuevas, Alejandro Flores-Morales, Laura

Ortiz-Hernández, Karen Flores-Trujillo, Fernando Ramos-Quintana, and Efraín Tovar-Sánchez. 2020. "Heavy Metal Bioaccumulation and Morphological Changes in *Vachellia Campechiana* (Fabaceae) Reveal Its Potential for Phytoextraction of Cr, Cu, and Pb in Mine Tailings." *Environmental Science and Pollution Research*, no. 1001. https://doi.org/10.1007/s11356-020-07730-7.

Siddiqui, Fauzia, Sunil Kumar Krishna, P. K. Tandon, and Sudhakar Srivastava. 2013. "Arsenic Accumulation in *Ocimum* Spp. and Its Effect on Growth and Oil Constituents." *Acta Physiologiae Plantarum* 35 (4): 1071–79. https://doi.org/10.1007/s11738-012-1145-1.

Silva, Deivisson Ferreira da, Patriciani Estela Cipriano, Ray Rodrigues de Souza, Matias Siueia Júnior, Rodrigo Fonseca da Silva, Valdemar Faquin, Maria Ligia de Souza Silva, and Luiz Roberto Guimarães Guilherme. 2020. "Anatomical and Physiological Characteristics of *Raphanus Sativus* L. Submitted to Different Selenium Sources and Forms Application." *Scientia Horticulturae* 260 (September 2019): 108839. https://doi.org/10.1016/j.scienta.2019.108839.

Silveira Rabelo, Ana Carolina, and Daniela Caldeira Costa. 2018. "A Review of Biological and Pharmacological Activities of Baccharis Trimera." *Chemico-Biological Interactions* 296: 65–75. https://doi.org/10.1016/j.cbi.2018.09.002.

Steinmann, Víctor W., Benjamin Van Ee, Paul E. Berry, and Jorge Gutiérrez. 2007. "The Systematic Position of *Cubanthus* and Other Shrubby Endemic Species of *Euphorbia* (Euphorbiaceae) in Cuba." *Anales Del Jardín Botánico de Madrid* 64 (2). https://doi.org/10.3989/ajbm.2007.v64.i2.167.

Stevanović, B., G. Dražić, G. Tomović, J. Šin Šinžar-Sekulić, Lj Melovski, I. Novović, and D. M. Marković. 2010. "Accumulation of Arsenic and Heavy Metals in Some Viola Species from an Abandoned Mine, Alchar, Republic of Macedonia (FYROM)." *Plant Biosystems* 144 (3): 644–55. https://doi.org/10.1080/11263504.2010.492597.

Texas A&M University. n.d. *Aggie Horticulture*. https://aggie-horticulture.tamu.edu/ornamentals/nativeshrubs/baccharisneglecta.htm.

Texas Tech University. n.d. *Plant and Soil Science*. https://www.depts.ttu.edu/pss/plantid/Pages/Information_Sheets/Herbaceous_Info/Miscanthus_sinensis.php.

Tiwari, Jaya, Poulomi Chakravarty, Pallavi Sharma, Ragini Sinha, Manoj Kumar, and Kuldeep Bauddh. 2021. "Phytoremediation: A Sustainable Method for Cleaning up the Contaminated Sites." In *Phytorestoration of Abandoned Mining and Oil Drilling Sites*, edited by Kuldeep Bauddh, John Korstad, and Pallavi Sharma, 3–32. Elsevier. https://doi.org/10.1016/B978-0-12-821200-4.00019-4.

Tonny, Tamanna Sharmin, Sonia Sultana, and Farhana Siddika. 2017. "Study on Medicinal Uses of Persicaria and Rumex Species of Polygonaceae Family Tamanna." *Journal of Pharmacognosy and Phytochemistry JPP* 6 (6): 587–89.

Torres-Avilez, Wendy Marisol, Flávia dos Santos Silva, Albuquerque, and Ulysses Paulino. 2007. "Cymbopogon Citratus (DC.) Stapf." *Journal of Al-Nahrain University Science* 10 (2): 52–63. https://doi.org/10.22401/jnus.10.2.41.

University of Winconsin Madison. 2020. *Mentha X Villosa DW.* Department of Botany. 2020. http://botit.botany.wisc.edu/Resources/Systematics/Magnoliophyta/Magnoliopsida/Lamiales/Lamiaceae or Labiatae/Mentha/Mentha X Villosa DW.jpg.html.

Vanderhoff, Ron. 2011. *Azolla Filiculoides.* Natural History of Orange Country. 2011. http://nathistoc.bio.uci.edu/plants/Ferns/Azolla filiculoides/Azolla filiculoides.htm.

Vassilev, Kiril, and Hristo Pedashenko. 2013. "Reports 141-146 - In: Vladimirov, V., Dane, F., Stevanović, V. & Tan, K. (Eds). New Floristic Records in the Balkans: 22." *Phytologia Balcanica* 19: 294. https://www.researchgate.net/publication/273145823_Reports_141-146_-_In_Vladimirov_V_Dane_F_Stevanovic_V_Tan_K_eds_New_floristic_records_in_the_Balkans_22.

Wahlert, Gregory A., Thomas Marcussen, Juliana de Paula-Souza, Min Feng, and Harvey E. Ballard. 2014. "A Phylogeny of the Violaceae (Malpighiales) Inferred from Plastid DNA Sequences: Implications for

Generic Diversity and Intrafamilial Classification." *Systematic Botany* 39 (1): 239–52. https://doi.org/10.1600/036364414x678008.

Williams, Paul. 2015. "*The Science Education Resource Center.*" 2015. https://serc.carleton.edu/genomics/units/brassica_photos.html.

Wu, Chuan, Bin Liao, Sheng Long Wang, Jun Zhang, and Jin Tian Li. 2010. "Pb and Zn Accumulation in a Cd-Hyperaccumulator (Viola Baoshanensis)." *International Journal of Phytoremediation* 12 (6): 574–85. https://doi.org/10.1080/15226510903353195.

Yakın Doğu Üniversitesi, Herbaryum Merkezi. 2020. *Sinapis Arvensis.* 2020. http://herbaryum.neu.edu.tr/resimler/brassicaceae-cruciferae/sinapis-arvensis/.

Yüzüncü Yıl Üniversitesi, Eğitim Fakültesi. 2020. "*Van Gölü Havzası Sanal Herbaryumu-Brassicaceae Raphanus.*" 2020. http://www.vanherbaryum.yyu.edu.tr/sifabit/turp/pages/Raphanus sativus_jpg.htm.

Zhao, Anqi, Lingyun Gao, Buqing Chen, and Liu Feng. 2019. "Phytoremediation Potential of Miscanthus Sinensis for Mercury-Polluted Sites and Its Impacts on Soil Microbial Community." *Environmental Science and Pollution Research* 26 (34): 34818–29. https://doi.org/10.1007/s11356-019-06563-3.

In: The Future of Phytoremediation
Editor: James C. Flores
ISBN: 978-1-53619-625-2

Chapter 2

THE INFLUENCE OF LIGHT QUALITY IN PLANTS: A NEW PERSPECTIVE FOR THE DECONTAMINATION OF ORGANIC COMPOUNDS

Murilo Ricardo Sigal Carriço, Mateus Cristofari Gayer, Marina Diaz Rodrigues, Cristine de Freitas Rodrigues, Maria Laura Videiro Schmitt, Elton Luiz Gasparotto Denardin and Rafael Roehrs
Laboratório de Análises Químicas
Ambientais e Toxicológicas (LAQAT),
Universidade Federal do Pampa, Uruguaiana, RS, Brazil

ABSTRACT

Phytoremediation is gaining more and more prominence in the scientific community, as an inexpensive and easy-to-implement technique for decontamination of water, soil, and air contaminated by inorganic (heavy metals) and organic compounds (pesticides, pharmaceuticals, personal care products). Since its inception, a wide range of laboratory and field studies have been developed to evaluate different plants, to accumulate, degrade or make these contaminants less toxic. Until then, some limitations to these perspectives have emerged, such as the low capacity of the plant to remedy certain compounds that, for the most part, have stressful effects on the metabolism of the plant, decreasing its efficiency and harming the health of the plant throughout the process. Thus, it would be possible to assume that improving the plant's response to the stress caused by these contaminants would have a major impact on improving phytoremediation. However, few studies aim to improve plant health as an alternative to increasing phytoremediation efficiency, both on a laboratory scale and in the field. When dealing specifically with some organic compounds (pesticides and metabolites). Recent studies show that oxidative stress induced by pesticides alters the cellular redox balance, altering the antioxidant response as the activity of cellular defense enzymes: SOD, CAT, APX, GPx, GST, and GR (directly related to xenobiotic detoxification), in addition to levels of secondary metabolism compounds in plants, with phenolics, flavonoids, anthocyanins, and terpenes. Therefore, these parameters can be used as tolerance criteria for the selection of plants with phytoremediation capacity. One way to stimulate the production of these defense mechanisms is through changes in the quality of light, it is known that the irradiation of some colors of light can stimulate the production of secondary metabolism compounds and detoxify enzymes such as those mentioned above, in addition to this to influence photosynthetic rate, root and stem growth, and flowering. Thus, in this chapter, we review some of the potentials described in the literature on ways to promote phytoremediation of toxic organic compounds and maintain plant health, so that it can better cope with stressful environments and thus decontaminate more efficiently. Besides, we bring some unprecedented results developed in our laboratory using photo-stimulation with LEDs as a phytosanity promoter, in the face of a simulated environment contaminated with the herbicide 2,4-D.

Keywords: defense mechanisms, secondary metabolism, antioxidants response, reactive oxygen species, light-emitting diodes, light quality

1. INTRODUCTION

The growing human activity has made the disposal of dangerous substances in nature more and more recurrent. Whether by agricultural, industrial, or urban activity, organic compounds are increasingly present in aquatic and terrestrial environments, which can negatively affect contaminated ecosystems and even human health. Becoming even more serious by population growth, which generates extreme needs, such as an increase in food production while increasing the use of drugs to promote health.

Currently, many contaminants are constantly released into nature, whether in the aquatic or terrestrial environment, triggering cumulative contamination that overloads ecosystems and generates significant environmental impacts (Daneshvar et al., 2012; Aubertheau et al., 2017). Thus, it is important to consider the level of the permanence of these compounds in the environment. Since, the use of pesticides is very common in agriculture, mainly due to the widespread use of pesticides to improve production and combat pests in agriculture. Besides, all these advances contribute to the increase of environmental contamination, especially in aquatic ecosystems, since leaching, percolation, and runoff from wastewater, make these environments the destination of many organic contaminants (Chrétien et al., 2017; Yang et al., 2017). Some studies show that molecular and cellular changes in aquatic populations due to the presence of contamination by organic substances, despite low concentrations, present possible chronic effects, promoting long-term physiological, behavioral, and population changes (Kidd et al, 2007; Kidd et al., 2012; Santos et al., 2010; Wrubleswski et al., 2018).

This is because many of these environmental contaminants have been present in aquatic ecosystems, or even in sources of drinking water consumed by the population due to deficiencies in sewage treatment or free runoff from crop waters (Kapsi et al., 2019; Sjerps et al., 2019; Styszko et al., 2021). In conventional water treatment plants (WTPs), the treatment of raw water with coagulation, flocculation, sedimentation, filtration, and chlorination may not be sufficient for the efficient removal of organic contaminants (Sultana et al., 2018). In sewage treatment plants (STPs), drugs and personal care products (PPCPs), in general, have a low rate of effluent removal and are one of the main routes of entry of PPCPs into aquatic ecosystems (Saxena et al., 2021; Subedi et al., 2017; Yang et al.,

2017). With this, it is essential that society seeks alternatives to reduce the rate of contaminants that reach the most varied ecosystems, or even, to reduce the permanence of these substances, to minimize the environmental and social impacts.

New alternatives for removing organic substances have emerged in recent years, with emphasis on phytoremediation. This strategy adds the ability of plants to absorb, metabolize and degrade organic substances from aquatic and terrestrial environments. With great importance for the remediation of organic compounds, enabling the use in plant metabolism and thus, minimizing the risk of producing toxic metabolites. Even phytoremediation is widely used to remove inorganic substances, such as heavy metals, but the risk of introducing plant biomass can lead to the accumulation of metals in the food chain (Kovacs & Szemmelveisz, 2017). However, the use of plants for the remediation of organic compounds has many attractions, in addition to being considered cheap. On the other hand, some studies show that phytoremediation still has some deficiencies, such as the high time required for remediation and the high toxicity of some contaminants to the plant (Abdel-Shafy & Mansour, 2018; Ekperusi et al., 2019). In this context, research has been developed worldwide with the objective of evaluating plants that have a greater capacity to phytoremediate environmental contaminants.

Some of the characteristics considered ideal in plants are those of hyperaccumulating or hyperdegrading the xenobiotics exposed to them. Directly linked to the ability of plants to deal with the toxic effects caused by exposure to harmful substances during phytoremediation. In a way, some xenobiotics inhibit photosynthetic activity, growth, cell division, and inhibition of antioxidant defenses, by the excessive production of reactive oxygen species (ROS), which leads to a wide range of metabolic problems in the plant. As a result, plants that can cope well in the face of stressful situations, caused by exposure to contaminants, can be considered promising to be used in phytoremediation studies.

Instead of looking for plants that have a higher phytoremediation rate or plants that would be more resistant to contaminants, an alternative was sought to make phytoremediation processes more efficient to improve the plant's ability to deal with a stressful situation. Thanks to positive stimuli that aim to improve the plant's metabolic defenses and stimulate growth, promoting plant health. In this context, some alternatives are already being studied, such as the inoculation of symbiotic bacteria and fungi that help the plant remedy (Deng & Cao, 2017; Wang, et al., 2019), genetic improvement to increase the accumulation of metals (Zhu et al., 2020), and physical and chemical additives, such as the improvement in light quality, the addition of fertilizers, chelators and nanomaterials, which influence the absorption, translocation, metabolic response and plant growth (Al Mahmud et al., 2018; Luo et al., 2020; Song et al., 2019; Wang et al., 2020). Thus, alternatives are emerging for phytoremediation to become increasingly efficient in its role of remedying contaminated environments, promoting an increase in the rate of absorption, degradation, and accumulation, and maintaining the health of the plant during this process. In this perspective, we bring some results that corroborated this trend, paving the way for the possible use of light stimuli in the phytoremediation of pesticides.

2. PHYTOREMEDIATION

Phytoremediation is defined as the efficient use of plants to remove, detoxify or immobilize environmental contaminants, recovering contaminated soil and water. The mechanism involves the absorption of pollutants by the roots, and accumulation in tissues and the body, leading to the decomposition and transformation of pollutants into less harmful forms (Ansari et al., 2020). Furthermore, phytoremediation is an advantageous ecological technique for the decontamination of environments, as it does not require physical-chemical or engineering processes such as excavation or incineration of the soil, in addition to having a lower cost about these techniques, with its possible execution *in situ* (Patra et al., 2020; Yang et al., 2005).

Also, this technique allows different mechanisms for the detoxification of water, soil, and air contaminated mainly with metals and organic compounds. All of these mechanisms occur after the xenobiotic absorption by the plant and are described in Table 1, for each phytoremediation technique (Yang et al., 2005).

The literature shows us that although they have reported that phytoremediation, in general, has attractive advantages when compared to other more complex processes for the remediation of contaminated waters and soils, this technique has mainly longer execution times. However, when we analyze each technique specifically (as described in Table 1), they all have specific care, added to unmatched pros and cons. This information is described in Table 2.

Table 1. Different phytoremediation strategies

Phytoremediation techniques	Mechanism of action	Medium treated
Phytoextraction	Direct accumulation of contaminants in plant shoots, with subsequent removal of plant shoots	Soil
Rhizofiltration (phytofiltration)	Absorb and adsorb pollutants to plant roots	Surface water and water pumped by the roots
Phytostabilization	Root exudates cause metals to precipitate and become less bioavailable	Groundwater, soil, mine tailings
Phytovolatilization	Plants evaporate certain volatile metallic and organic ions	Soil, groundwater
Phytegradation (plant-assisted bioremediation)	Microbial degradation in the rhizosphere region	Groundwater in the rhizosphere and soil
Phytotransformation	Absorption of plants by organic contaminants and degradation	Surface and groundwater
Removal of airborne contaminants	Uptake of various volatile organics by the leaves	Air

Source: Yang et al., (2005).

Table 2. Advantages and disadvantages of each phytoremediation mechanism

Phytoremediation mechanism	Advantages	Disadvantages
Phytoextraction	1. The cost is genuinely cheap, in contrast to the usual strategies. 2. The measure of the residual material that must be discarded decreases up to 95% 3. Sometimes the contaminant can be reused. 4. Contaminant expelled from the soil at all times.	1. Extension of the metal "seated" within the roots. 2. Rate of absorption of metals by the roots.
Phytostabilization	1. No transfer of risk material/biomass is necessary 2. Extremely successful when rapid immobilization is expected to save groundwater and surface water	1. Permanence of contaminants in the soil 2. Use of extensive treatment/soil reviews 3. Mandatory verification required
Phytodegradation	1. Both financially and naturally neighborly	1. Requires more than one developing season to be productive 2. Soil must be under 3 ft inside and out and groundwater inside 10 ft of the surface 3. Contaminants may in any case reemerge the natural way of life through creatures or creepy crawlies that eat plant material

Table 2. (Continued)

Phytoremediation mechanism	**Advantages**	**Disadvantages**
Rhizodegradation	1. In situ work on coming about in no unsettling influence 2. No evacuation of sullied materials 3. Finish mineralization of the contaminant can happen 4. Low establishment and upkeep cost	1. Advancement of broad root zone required - takes time 2. Root profundity restricted due to physical structure of soil 3. Natural issue from plant might be utilized as a C source rather than contaminant ⇒ diminish measure of contaminant biodegradation
Phytovolatilization	1. The contaminant, mercuric particle, might be changed into a less lethal substance (i.e., essential Hg)	1. The mercury discharged into the air is probably going to be reused by precipitation and afterward re-saved back into lakes and seas, rehashing the generation of methyl–mercury by anaerobic microscopic organisms.
Rhizofiltration	1. Capacity to utilize both earthly and oceanic plants for either in situ also, *ex situ* applications. 2. Contaminants do not be translocated into shoots.	1. Steady need to alter pH 2. Plants may initially need to be developed in nursery/nursery 3. There is occasional collecting and plant transfer 4. Tank configuration ought to be very much built

Source: Jeevanantham et al., (2019).

Phytodegradation consists of the process of degradation of contaminants by the plant through metabolic processes within the plant itself. This process occurs through enzymes such as oxygenases and nitro-reductases that are produced by plants and catalyze the degradation of xenobiotics, this process is ideal for organic pollutants that can be used as substrates for the metabolic needs of the plant (Jeevanantham et al., 2019; Rosa, 2013).

Phytoextraction or phytoaccumulation uses hyperaccumulating plants capable of extracting contaminants from water or soil and accumulating them in their tissues. These plants are capable of accumulating various inorganic elements at levels up to a hundred times greater than other species, growing under the same conditions (Rosa, 2013). The plants used for phytoextraction must-have characteristics of fast growth, high biomass, easy cultivation, easy harvesting, highly tolerant to stress, and inedible to prevent the entry of harmful substances into the food chain (Patra et al., 2019; Patra et al., 2020).

Phytostabilization is the process in which soil and water contaminants are immobilized by the use of plants, by adsorption, accumulation in the roots, or precipitation in the root zone, thereby preventing their dispersion (Jeevanantham et al., 2019; Tangahu et al., 2011). The immobilization of xenobiotics provided by this technique, in addition to reducing the leaching of toxic metals into groundwater and the agricultural area, decreases their concentration in the food chain (Patra et al., 2020).

In phytovolatilization, the mechanism consists of removing contaminants through the leaves of plants into the atmosphere through the transpiration of the plant (Tangahu et al., 2011). Contaminants become modified in water-soluble and non-toxic form during transport from the root to the leaves along the plant until compartmentalization in the vacuoles, and then the contaminants are volatilized (Jeevanantham et al., 2019). During the rhizodegradation, there is the process of biotransformation or breakdown of organic contaminants present in the soil, carried out in the soil area around the roots of the plants (rhizosphere). This mechanism involves the association of microorganisms (fungi and symbiotic bacteria) with plants (Ramborger et al., 2017). Roots change soil conditions by increasing aeration, adjusting moisture and sugars, alcohols and organic acids act as a carbon source for soil microbes, thus increasing microbial activity. This technique is also called enhanced rhizosphere biodegradation, phytostimulation, or plant-assisted bioremediation (Jeevanantham et al., 2019). Table 3 shows some of the enzymes that perform this process.

Rhizofiltration involves the adsorption and precipitation of metallic contaminants on the growth substrate around the root zones. In this process, contamination of groundwater, surface water, occurs through the roots of the plant (Ansari et al., 2020; Dushenkov et al., 1995). Thus, the mechanism of rhizofiltration is based on the fact that contaminants present in polluted water or soil, are absorbed or adsorbed on the roots of plants. This mechanism is similar to that of phytoextraction, however, it is performed mainly in the soil area in the plant's roots (Jeevanantham et al., 2019).

Table 3. Vegetable and microbial enzymes with a role in the degradation of organic compounds

Enzyme Family	**Catalytic Action**	**Examples of Known Sources**
Laccase	Degradation of various aromatic compounds	*Populus* spp.
Dioxygenase	Degradation of various aromatic compounds.	*Medicago sativa*
Nitrilase	Cleaves aromatic and aliphatic nitrile cyanide groups.	*Armoracia rusticana. Medicago sativa*
Nitroreductase	Reduces nitro groups of nitroaromatic compounds, removes nitrogen from cyclic structures.	*Salix* spp. *Populus* spp.
Phosphatase	Cleaves phosphate groups from organophosphates.	*Spirodela polyrhiza*
Cytochrome P450 monooxygenase	Hydroxylation of aliphatic and aromatic hydrocarbons.	Majority of aerobic bacteria, all fungi and plants.

Source: Reorganized data from Nahler et al., (2009).

3. Plant Detoxification Process

Plants play a fundamental role in the removal of toxic compounds from the environment, when dealing with organic compounds, the removal process involves several steps from which many details have yet to be known and elucidated. These compounds, either by direct or indirect action of the plant, can be stabilized or degraded in the plant's rhizosphere, absorbed and accumulated in the roots, or even transported by the plant, until they are volatilized or degraded by its metabolism (Collins et al., 2006).

Many organic compounds because they are xenobiotics do not have specific transporters in plants, and they move by diffusion between tissues due to the potential gradient created by perspiration. Also, the translocation of these compounds varies according to their physical-chemical characteristics, such as water solubility, log Kow, pKa in addition to the plant's own characteristics. However, absorption has the primary control over the translocation of xenobiotics, the metabolism influences both the uptake of xenobiotics and its phytotoxic action, making the compound less or more active (Dordio & Carvalho, 2011; Kvesitadze et al., 2006).

Once organic xenobiotics are absorbed by plant roots and translocated to different organs, they penetrate plant cells and are susceptible to metabolic transformations in the plant, which can be partially or completely degraded. And with that, it can be transformed into less toxic compounds and linked to plant tissues (Dordio & Carvalho, 2011; Kvesitadze et al., 2006). In general, this is seen as a detoxification process, very similar to that of mammals. Once an organic xenobiotic is absorbed and translocated, it undergoes one or more phases of metabolic transformation. Three possible phases of metabolic transformation of organic compounds can be identified in higher plants, such as Phase I, II and III (Sandermann, 1994), the passage mechanism of xenobiotics can be seen in Figure 1A.

Phase I is known for a transformation or conjugation phase, in which lipophilic compounds (apolar) acquire a hydrophilic group (hydroxyl, amine, carboxyl, sulfhydryl), catalyzed by enzymes. In this phase, the oxidative metabolism of xenobiotics is carried out mainly by the cytochrome P450 monooxygenase, that is, oxidative processes of the bioactivation of the compounds in chemically reactive compounds. Favoring solubility in water, which also causes an increased affinity for enzymes that catalyze the transformation to Phase II (Dordio & Carvalho, 2011).

Phase I reactions are the first step necessary to make a molecule into a less toxic xenobiotic, these reactions modify the molecule for Phase II reactions. However, if the molecule already has a functional group suitable for Phase II metabolism, the compound can be used directly in this step (Dordio & Carvalho, 2011).

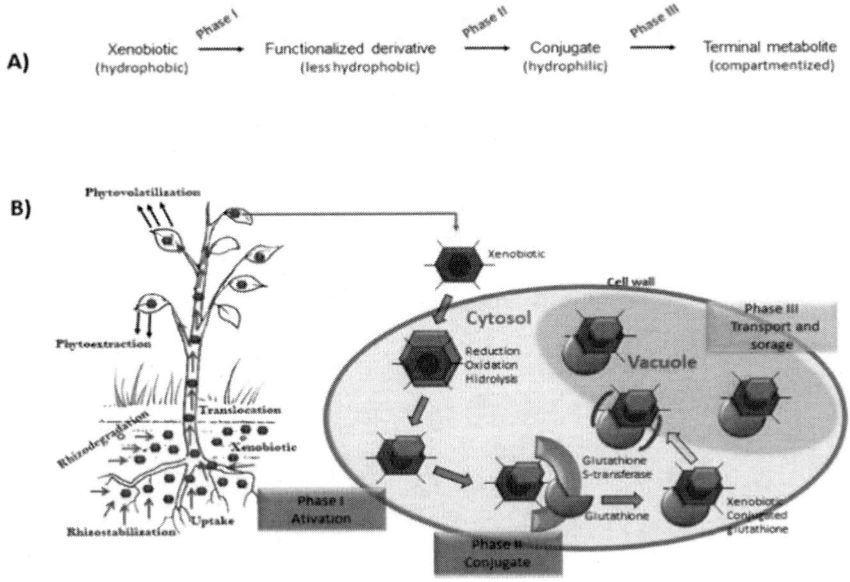

Figure 1. (A) Simplified scheme of xenobiotic degradation in plants. E (B) Typical plant attenuation mechanism against xenobiotics. Plant detoxification usually involves conversion or enzymatic modification (phase I) followed by conjugation (phase II) followed by active sequestration (phase III). Xenobiotics can be stabilized, degraded in the rhizosphere, adsorbed, or accumulated in the roots and transported to the aerial part, volatilized or degraded inside the tissue.

Phase II, called the conjugation phase, is characterized by the conjugation of xenobiotics to endogenous molecules such as proteins, peptides, amino acids, organic acids, sugars, pectins, and lignins (Coleman et al., 1997; Dietz & Schnoor, 2001; Korte et al., 2000). This step has enzymes responsible for catalyzing conjugations, such as glutathione-S-transferases, glucosyltransferase, and N-malonyl transferases (Eapen et al., 2007). The conjugation of these molecules increases the molecular weight and significantly reduces the biological activity of these compounds, obtaining a less toxic metabolic product than those from Phase I (Dordio & Carvalho, 2011).

Phase III reactions are exclusive to plants, as they do not excrete xenobiotics like animals, therefore, they need to somehow remove the compounds from their own system, using vacuolar transporters for this. These transporters are the main enzymes involved in Phase III that consume ATP, to carry out the passage of compounds to the vacuoles, where they are stored (Dordio & Carvalho, 2011; Eapen et al., 2007).

Xenobiotics are then compartmentalized in vacuoles or bind to components of the cell wall, such as lignin and hemicellulose. Vacuolar compartmentalization is an important step in the detoxification of xenobiotic conjugates in plants since the conjugated compounds are removed from vulnerable sites in the cytosol and transported to vacuoles or apoplast by transporters connected to the tonoplast membrane, where they do not interfere with cell metabolism (Coleman et al., 1997). The simplified path of this set of mechanisms can be seen in Figure 1A, and the general mechanism can be seen in Figure 1B.

4. Formations of Reactive Species and Oxidative Stress

The evolution of aerobic metabolic processes, such as respiration and photosynthesis, inevitably leads to the production of ROS in the mitochondria, chloroplast, and peroxisome. Under normal growing conditions, ROS are naturally produced in low quantities in organelles and by the super-reduced photosynthetic electron transport chain (You & Chan, 2015). In addition to organelles, the plasma membrane together with the apoplast is the main site of ROS generation in response to endogenous signals and exogenous environmental stimuli (You & Chan, 2015).

Therefore, the photosynthetic apparatus is one of the main stress targets in plants. Most of the metabolic responses induced by stress conditions have consequences on the plant's ability to maintain photosynthesis efficiently (Rmiki et al., 1999). Thus, changes in the levels of chlorophyll and carotenoids are correlated with stress levels in the plant. For plants in general, the composition of carotenoids tends to increase to provide protection against the formation of ROS. At the same time, there is a tendency towards a decrease in chlorophyll levels. This trend has been observed previously by the intoxication of several classes of xenobiotics. Thus, physiological responses such as chlorosis, necrosis and growth regulation are easily obtained, such as quantitative physiological parameters, such as relative growth or levels of photosynthetic pigments, which also evaluate responses toxicological effects of the plant (Dordio & Carvalho, 2011).

Biochemical changes induced by the presence of organic xenobiotics lead to the production of excess ROS. These chemical species are reduced forms of atmospheric oxygen (O_2) generating superoxide radicals ($O_2^{\bullet -}$), hydrogen peroxide (H_2O_2) or $^3O_2$, which is dismuted in water and hydroxyl radicals ($OH^{\bullet}$), as seen in detail in Figure 2 (Apel & Hirt, 2004; Dordio e Carvalho, 2011; Noctor et al., 2014).

Under normal conditions, ROS are easily eliminated by different antioxidant defense components, usually confined to specific locations (Apel & Hirt, 2004). However, its overproduction can be triggered by abiotic stressors, such as exposure to xenobiotics, such as pesticides. When exposed to these compounds, plants activate pathways to metabolize compounds that are capable of triggering large amounts of ROS and can alter homeostasis between the production and elimination of ROS. These disorders can quickly increase the levels of intracellular ROS and consequently impair cell viability (Figure 3) (Dordio & Carvalho, 2011).

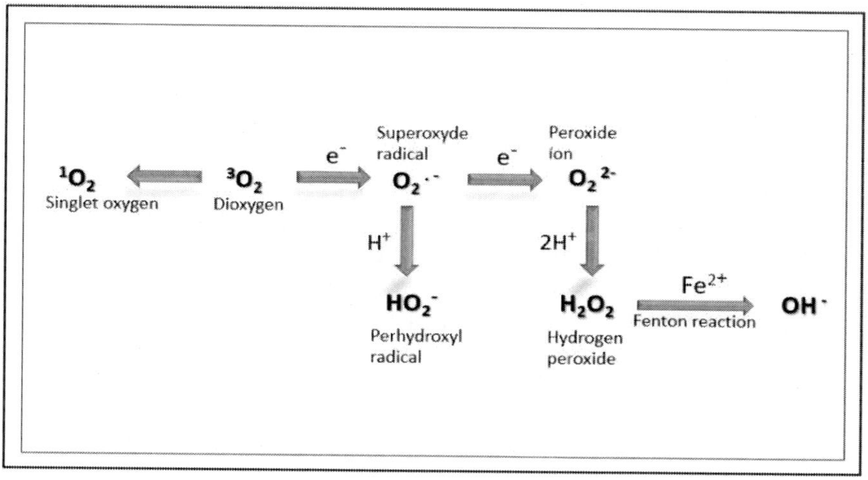

Figure 2. Generation of Reactive Oxygen Species.

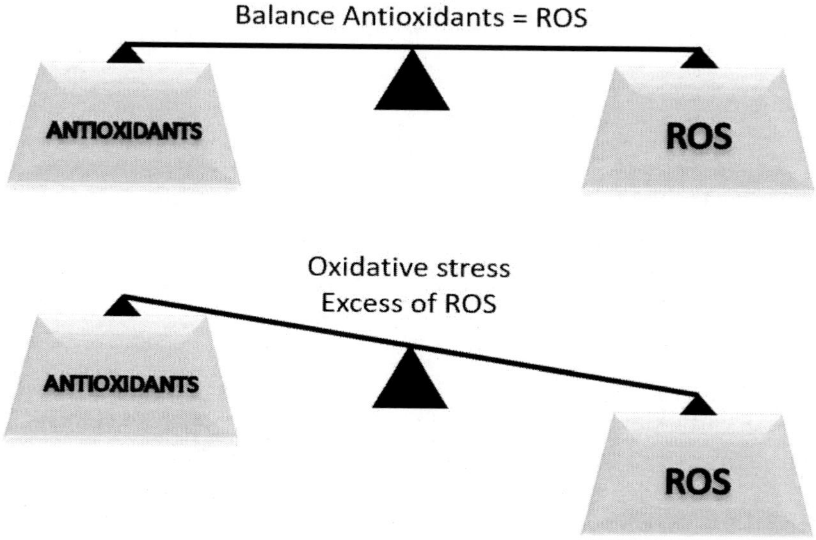

Figure 3. Balance among Antioxidants and Reactive Oxygen Species.

ROS are capable of unrestricted oxidation of a large number of cellular components, and if not controlled, lead to damage to biomolecules such as membrane lipid peroxidation, protein oxidation, enzyme inhibition and DNA and RNA damage (Mittler et al., 2002). Lipid peroxidation starts in the presence of ROS, as plants are often exposed to stressful environments. Under these conditions, ROS accumulates and its toxicity can be exerted through its interaction with biomolecules such as lipids, proteins and nucleic acids, sometimes resulting in cell death, and consequently, a limitation in biomass and yield production (Alche, 2019).

The lipid peroxidation process results mainly from the action of $OH^{\bullet}$, which when sequestering an H^+ of a polyunsaturated fatty acid belonging to the phospholipid bilayer of cell membranes, promotes the destabilization of this forming an instability region, the alkyl radical ($R^{\bullet}$), possessing an unpaired electron. This, in turn, reacts with an O_2 molecule and can form two other radicals, alkoxy ($RO^{\bullet}$) or peroxyl ($ROO^{\bullet}$), which tend to stabilize by removing an H^+ molecule and an electron belonging to the adjacent phospholipids, forming lipid hydroperoxide (ROOH) and leading to the formation of a new $R^{\bullet}$, promoting a chain reaction (Amaral et al., 2018).

Because of this, ROS levels must be precisely controlled (kept in a small range), and for this, plant cells have mechanisms to monitor and eliminate excessive amounts of ROS. Thus, the tolerance of plants to toxic agents is related to their ability to deal with ROS (Dordio & Carvalho, 2011).

This fine regulation of ROS levels is more important, considering that these molecules have the function of signaling for the activation of the stress response and its defense pathways. Therefore, ROS can be seen as cellular indicators of stress, as well as secondary messengers in the signal transduction pathway to stress response, as well as the measurement of the activity of antioxidant enzymes, which make it possible to detoxify ROS (Apel & Hirt, 2004; Mittler, 2002).

5. ROS Detoxification Mechanisms

Two different mechanisms are involved in the modulation of ROS levels, one that performs fine modulation of ROS levels for signaling purposes. And another that performs the detoxification of excess ROS, especially during oxidative stress. These mechanisms exist in all plants and are divided into enzymatic and non-enzymatic capable of neutralizing the cytotoxicity of ROS. The cellular antioxidant defense system begins with an enzymatic cascade, but also involves non-enzymatic components. The main enzymes responsible for ROS detoxification include superoxide dismutase (SOD), ascorbate peroxidase (APX), glutathione peroxidase (GPX) and catalase (CAT), glutathione (GSH), β-carotene and α-tocopherol (Figure 4). In addition to non-enzymatic systems, produced by secondary metabolism, such as flavanones, anthocyanins, carotenoids and ascorbic acid (AsA) (Apel & Hirt, 2004; Dordio & Carvalho, 2011; Kasote et al., 2015; Mitter, 2002).

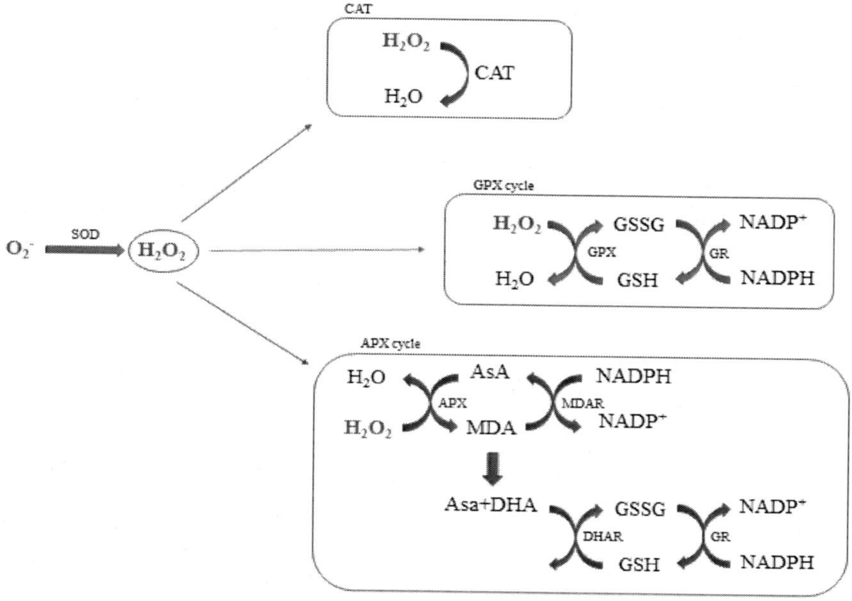

Figure 4. Detoxification pathways of reactive oxygen species by Catalase (CAT), the Glutathione Peroxidase cycle (GPX) and the Ascorbate Peroxidase cycle (APX). Monodehydroascorbate (MDA), ascorbic acid (AsA), dehydroascorbate (DHA), dehydroascorbate reductase (DHAR), Glutathione Reductase (GR). Superoxide dismutase (SOD), MDA reductase (MDAR), oxidized glutathione (GSSG), glutathione (GSH).

Plants synthesize low molecular weight antioxidants, such as glutathione and ascorbate within the chloroplast and cytosol stroma, functioning as redox buffers that interact with various cellular components and influence plant growth and development (Alscher et al., 2006). Enabling, modulating processes of mitosis and cell elongation to senescence and death, as these antioxidants can influence gene expression associated with biotic and abiotic stress responses to maximize defense (Kasote et al., 2015). The enzymes involved in this process are located in different cell compartments, and their expression is controlled by environmental and endogenous stimuli, depending on the need for ROS removal, as shown in the mechanisms of Figure 4.

Several genes encoding SOD and APX in plants, with different isoforms targeting the chloroplast, mitochondria, peroxisomes, apoplast and cytosol. GPX is located in the cytosol, CAT is located mainly in peroxisomes. In this way, the extent of oxidative stress in the cell is determined by the concentrations of superoxide, hydrogen peroxide and hydroxyl radicals. Therefore, the balance of SOD, APX and CAT activities will be crucial to suppress levels of toxic ROS in the cell (Garg & Manchada, 2009).

SOD is the first line of defense against ROS, dismuting superoxide (O_2^-) in H_2O_2, so that enzymes such as APX, GPX and CAT then detoxify H2O2. APX is considered an active line of defense for neutralizing excess production and possible negative effects of H_2O_2 (Schneider et al., 2020). CAT, present in the peroxisome, is responsible for inactivating the H_2O_2 formed during the conversion of glycolate to glyoxalate, which occurs during photorespiration, where the peroxide is converted by the enzyme to H_2O and O_2 (Igamberdiev & Lea, 2002). On the other hand, APX is an important enzyme in the defense of photosynthetic tissues against photooxidative stress, this enzyme is also found in the cytosol of non-photosynthetic cells, reducing ROS levels requiring an ascorbate and glutathione (GSH) regeneration system, (Figure 4) (Dordio & Carvalho, 2011; Garg & Manchada, 2009). However, the balance of SOD, APX and CAT activities will be crucial to suppress levels of toxic ROS in the cell (Garg & Manchada, 2009).

As already mentioned, the enzymes involved in this process are located in different cell compartments, and their expression is controlled by environmental and endogenous stimuli, depending on the need for ROS removal. Different isoforms of SOD and APX are directed to the chloroplast, mitochondria, peroxisomes, apoplast and cytosol. GPX is located in the cytosol, CAT is located mainly in peroxisomes. Thus, the extent of oxidative stress in the cell is determined by the concentrations of superoxide, hydrogen peroxide and hydroxyl radicals (Garg & Manchada, 2009).

The detoxification of H_2O_2 into H_2O by APX occurs by oxidation of ascorbate to monodehydroascorbate (MDA), being regenerated by MDA reductase (MDAR) using NADPH. MDA spontaneously dismutes itself in ascorbate and dehydroascorbate (DHA). Its regeneration is mediated by dehydroascorbate reductase (DHAR) conducted by the oxidation of GSH to oxidized glutathione (GSSG). Finally, glutathione reductase (GR) regenerates GSH from GSSG using NADPH. Like APX, GPX also detoxifies H_2O_2 into H_2O, but uses GSH directly as a reducing agent. The GPX cycle is closed by the regeneration of GSH from GSSG by GR using NADPH (Figure 4), (Dordio & Carvalho, 2011; Garg & Manchada, 2009; Mittler, 2002).

Recent studies list antioxidant responses as a measure of tolerance for plant selection for phytoremediation purposes, considering that although pesticides and xenobiotics act in different ways, oxidative stress is a convergent point in the mechanisms of toxicity (Bashir et al., 2007; Liu et al., 2019; Mitton et al., 2018; Vafaei et al., 2013). In addition, the behavior of antioxidant defenses depends on the chemical and physical properties of the compound, the intensity of exposure and the level of stress, as well as the internal functioning, which may vary according to the species of plants.

Liu et al., (2019) noted an increase in the activity of antioxidant enzymes using *Salvinia natans* during the phytoremediation of glyphosate and Cu alone and in the mixture. The authors observed an increase in the POD activity in the treatments with toxic agents (isolated and in the mixture) in relation to the control, however without a dose-dependence relationship. CAT also showed a gradual increase in relation to the control for all treatments of low concentrations of toxic agents (1 and 5 mg.L^{-1}), concentrations above these (up to 75 mg.L^{-1}) reduced their activity at the control level. . APX's activities in treatments showed a profile similar to that seen for CAT. It is well known that H_2O_2 produced in plants is mainly eliminated by enzymatic H_2O_2 scavengers alone or in combination, including CAT, SOD and APX. In general, the study showed that the content of H_2O_2 and MDA increased significantly when *S. natans* was exposed to single or combined stresses of glyphosate and Cu, added to other phytochemical parameters this result indicated the presence of oxidative stress in *S. natans* and the ability of the aforementioned enzymes to neutralize ROS and the cascade of reactions from these species to intermediate concentrations of toxic agents.

Santos et al., (2020) evaluated the phytoremediation and morphological changes of glyphosate (standard and commercial formulation) and suggested that the greatest efficacy of *Salvinia biloba* in remedying glyphosate is in the range of 1–30 mg.L^{-1}, as for this concentration range, efficient herbicide removal was observed, which can be correlated with the lowest phytotoxicity found. Bearing in mind that the activity of CAT, SOD and POD had a reduction in treatments with more than 30 mg.L^{-1} of glyphosate for both formulas, added to a slight reduction in the removal of the herbicide in the highest concentration tested (240 mg.L^{-1}). Demonstrating the correlation between plant health and its phytoremediation effect.

The results of Ramirez-Sondoval et al., (2011) strengthen this hypothesis because just like glyphosate, endosulfan is a pesticide that we assume promotes the alteration of defense mechanisms in plants so that the correlation between the lower levels of oxidative stress and the improvement in phytoremediation performance is remarkable. The authors evaluated the phytoremediation effect of *Ocimum basilicum L.* and *Ocimum Minimum L.* against endosulfan and noted that *O. basilicum* can be a suitable candidate for phytoremediation of soils polluted with endosulfan, taking into account the 37% reduction in the concentration of the herbicide. In addition, its regular growth rate and levels of lipid peroxidation were lower, when compared to treatments with *O. Minimum*

L., which presented higher levels of lipid peroxidation, lower growth rate and amounts of phytoremediation herbicide.

Several studies involving the viability of plants for phytoremediation of DDT (dichlorodiphenyltrichloroethane) and its commonly generated metabolites, (DDD and DDE) follow the same trends as the studies previously cited. Mitton et al., (2014) determined the potential for using the antioxidant response in assessing the tolerance of tomatoes, sunflowers, soybeans and alfalfa for species selection for phytoremediation purposes. The results of this study demonstrate that all plants store DDE and large amounts of DDE mainly in the roots in the first 15 days of treatment, when compared to the analysis in 60 days, especially the absorption of tomatoes. While the accumulation in the aerial parts stands out only the alfalfa. When we relate these data to the lipid peroxidation and growth of the specimens, we notice that this lower amount of DDT and DDE (in the leaves and roots) after the 60 days of treatment for all plants is correlated with the increase in lipid peroxidation, and the growth rate when compared to the values found in 15 days. However, DDT levels and soil dehydrogenase activity in soil fractions may be linked to variations in microbial communities due to different types and amounts of root exudates. This occurs because, at 15 days, the greatest microbial activity may be responsible for the metabolism of DDD in tomatoes and soybeans under anaerobic conditions, generating products not evaluated in this study.

In this perspective, the authors observed that, based on its greater accumulation potential, without significant phytotoxicity effects and increased microbial activity in the soil, tomatoes are considered the best candidate for phytoremediation. The study by Mitton et al., (2018) corroborates this hypothesis, because, after exposure to DDTs, tomatoes showed an increase in GR and GPX activity in the stems and leaves, respectively, and a decrease in GST activity in the roots, which in general shows a response against toxicity derived from the absorption of DDT, responsible for the high levels of pesticides in the roots.

Among the non-enzymatic defense mechanisms, phenolic compounds are important in the processes of detoxification and response to oxidative stress, are secondary metabolites present in all plants, and can play an important role in both non-enzymatic and enzymatic mechanisms (guaiacol peroxidase). In addition, phenolic compounds have an antioxidant action that allows them to directly neutralize free radicals, making them an important tool against the spread of oxidative stress (Dordio & Carvalho, 2011).

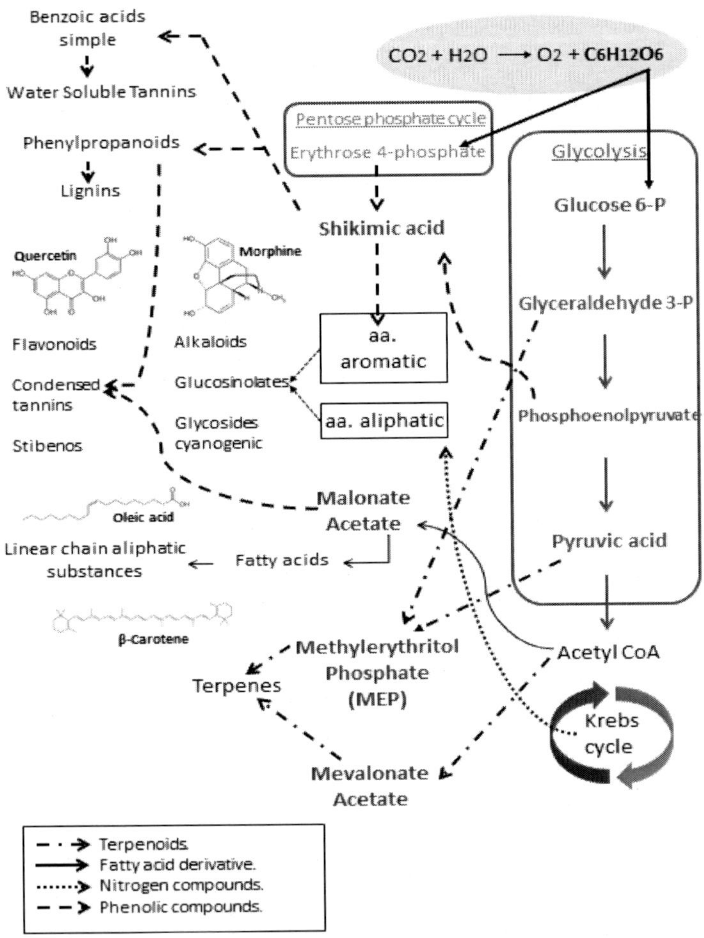

Figure 5. Biosynthetic cycle of secondary metabolites. The arrows indicate the formations of the main classes of secondary metabolites. The structures represent examples of metabolites synthesized via secondary plant metabolism. Bold are the main pathways that generate these compounds.

Secondary metabolites are products of plant metabolism, not directly related to the so-called primary processes for the plant, such as photosynthesis and respiration. Thanks to this, this set of biosynthesis pathways has been called secondary metabolism.

The secondary metabolites are extremely diverse (with more than 50 thousand compounds), their metabolic routes are not general, since the biochemical reactions do not occur independently of the same producer, which originates extremely diverse metabolites. Many secondary metabolites are formed by sequences of reactions analogous to those of the primary metabolism, the origin of the secondary metabolites can be summarized from the glucose metabolism, via main intermediates: chiquimic acid, malonate acetate, mevalonate acetate and metileritritolfosphate, as can be seen in Figure 5. Through them the four main groups are formed: derivatives of fatty acids, phenolic compounds, terpenes and nitrogenates (Meyer et al., 2013).

Despite this, the compounds originating from these pathways are vital for plants, presenting defense functions against herbivores and pathogens, allelopathy, protection against UV rays, pollution, the attraction of pollinators and dispersers, regulation of metabolism and molecular signaling. Secondary metabolism compounds are also responsible for mediating plant interactions with the environment, responsible for color and aroma. The secondary metabolism compounds due to their constitution, in a joint or isolated form (active principle), attribute the medicinal and/or toxic activity to these plants (Meyer et al., 2013; Santos et al., 2007).

Phenolic compounds are substances widely distributed in nature, more than 8,000 phenolic compounds have already been detected in plants. These compounds have in their structure, several characteristic benzene groups, having hydroxyl groups as substitutes. The hydrogen atoms of the adjacent hydroxyl groups (ortho-diphenols), located in various positions of the rings A, B and C, the double bonds of the benzene rings and the double bond of the oxo function (-C=O) of some flavonoid molecules guarantee these compounds their high antioxidant activity (Silva et al., 2010). The importance of phenolic compounds is directly related to physiological activity, the ability to capture ROS, chelate metal ions, inhibit nitrosation, inhibit the potential for auto-oxidation and have the ability to modulate certain active cellular enzymes (Port's, 2011).

Phenolic compounds can be synthesized via the shikimic acid pathway, via the malonate acetate pathway, or both, such as flavonoids that have mixed biosynthesis: the chikimic acid pathway provides a C6 - C3 ring while the pathogen malonate pathway offers the other ring C6 (Figure 6). These compounds can be found in free form (aglycone), linked to sugars (glycosides), proteins, terpenes, among others. Among the phenolic substances we have: phenolic acids, quinones, phenylpropanoids, coumarins, flavonoids, tannins and lignins (Meyer et al., 2013). Examples of phenolic compounds and their precursors are shown in Figure 6.

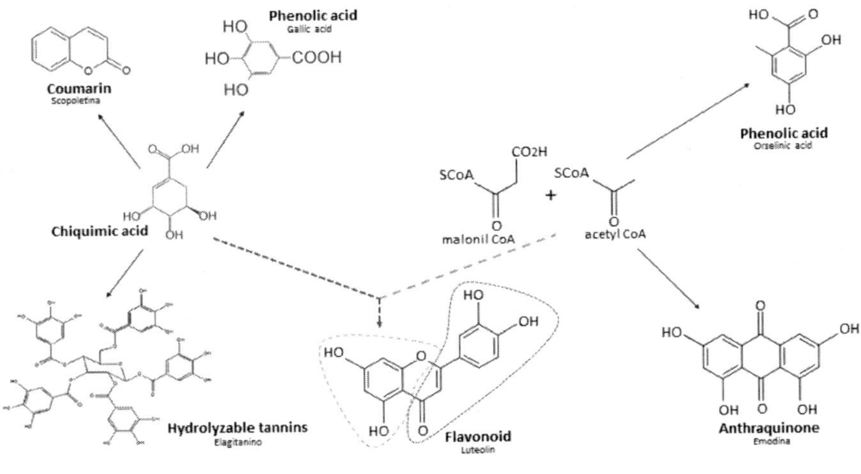

Figure 6. Examples of phenolic substances and their precursors, red lines indicate groupings referring to the malonate acetate pathway.

Among these compounds, flavonoids are of exceptional interest and represent one of the most diverse and largest groups of natural plant products. Flavonoids are a diverse group of secondary metabolites that consist of more than 10,000 structures present in various natural resources (Baskar et al., 2018). Much of the growing interest in flavonoids is due to their beneficial properties in relation to human health. However, these compounds also play important roles in the plants themselves, especially as pigments and defense compounds (Ververidis et al., 2007). In plants, flavonoids are involved in a number of processes, including plant-pathogen interactions, pollination, light screening, seed development and allelopathy (Winkel-Shirley, 2001).

Flavonoids and phenolic acids are molecules characterized by having at least one aromatic ring with one or more hydroxyl groups attached. Flavonoids and phenolic acids are composed of a wide range of structures and can be classified based on the number and arrangement of their carbon atoms into flavonoids (flavonoids, flavones, flavan-3-ols, anthocyanidins, flavanones, isoflavones and others) and non-flavonoids (phenolic acids, hydroxycinnamic acids, hydroxybenzoic acids, stilbenes and others) (Gutiérrez-Grijalva, 2018).

Flavonoids are the largest class of phenolics and comprise a group of compounds widely distributed in fruits and vegetables, presenting in several variations. These groups have a wide range of biological effects such as antioxidant, anti-inflammatory and anti-tumor activity and inhibition of collagen damage (Pereira & Cardoso, 2012).

Thus, flavonoids have been suggested as antioxidants, protecting plants from oxidative stress, as they are one of the main non-enzymatic antioxidants produced in the stressed plant and are involved in suppressing the generation of ROS. Detoxification of ROS in plants is accomplished through highly specialized elimination systems, including antioxidants and antioxidant enzymes. The prevention of the generation of ROS by flavonoids (antioxidants) is achieved through suppression of singlet oxygen; inhibition of ROS-generating enzymes including cyclooxygenase, lipoxygenase, monooxygenase and xanthine oxidase; the chelation of transition metal ions, such as Fe and Cu; and the recycling of other antioxidants (Baskar et al., 2018; Mierziak et al., 2014).

There are few studies in the literature that evaluate the influence of flavonoids as defense mechanisms during the phytoremediation of organic compounds. Ramborger et al., (2017a) evaluated whether there was variation in the antioxidant activity and in the composition of phenolic and flavonoid compounds due to oxidative stress caused by 2,4-

dichlorophenoxyacetic acid (2,4-D) in *Plectranthus neochilus* during phytoremediation in water and soil. The study showed that the specimens in the soil suffered damage from the herbicide during the 15 days of treatment, as it is notorious for the reduction of the levels of total flavonoids, total phenolics and consequently of the antioxidant potential. This fact occurs due to the pro-oxidant condition caused by exposure to the herbicide, followed by damage to the antioxidant system. This is confirmed because the specimens of *P. neochilus* added to the water treatments also had a reduction of these parameters in 30 days of treatment, however after that period the seedlings were replaced by new ones, and in 60 days the plants had a small increase in the total flavonoids, polyphenols and antioxidant capacity. This was possibly due to the lower concentrations of 2,4-D in the medium, leading to less toxicity in the specimens of *P. neochilus*. The changes in phenolic acids were also a factor related to stress, considering that there was a decrease in caffeic acid and an increase in coumaric and ferulic acids in relation to the control group (without 2,4-D). In the remaining 30 days with the new seedlings there was a decrease in coumaric acid and an increase in caffeic and ferulic acids. Which allows us to suppose a possible modulation of oxidative stress through these compounds.

To a diversity of secondary metabolism compounds whose changes are correlated with the action of toxic agents. Like anthocyanins, for example, Duran and Coskun (2015) investigated the effect of the insecticide deltamethrin on the biological properties of corn (*Zea mays L. saccharata Sturt*). The authors noted an increase in anthocyanins as the concentration of the pesticide increased. In addition, it was seen a decrease in chlorophyll a and b, total chlorophyll and carotenoids as the deltamethrin concentration increased, in addition to the dose-dependent increase in proline levels, thus configuring plant damage related to the increase in the deltamethrin concentration. Anthocyanins are flavonoid compounds responsible for protecting both chlorophyll pigments (a and b) from photoinhibition, photo-oxidative effects of different factors of environmental stress.

This behavior in increasing the flavonoid content as antioxidant protection can also be observed in phytoremediation studies of heavy metals that evaluate the influence of these metabolites in the phytoremediation process. Hussain et al., (2021) evaluated the application of ZnO nanoparticles (NPs) in *Persicaria hydropiper L.* for efficient restoration of water polluted with Pb through hydroponic studies. Where Pb significantly inhibited the growth of seedlings and NPs, ZnO NPs relieve stress induced by Pb, promoting plant growth and improving the content of chlorophyll and carotenoid. In addition, oxidative stress was mitigated with the use of ZnO NPs through increased production of free proline, phenolics, flavonoids, in addition to the activation of antioxidant enzymes. It is interesting to note that the same correlation between increased antioxidant defenses and phytoremediation potential applies in metal phytoremediation studies, since ZnO NPs also provided greater Pb absorption in roots and leaves than in treatment without NPs.

In this perspective, the work brought up in this review allows us to affirm that flavonoids are defense mechanisms that play several roles in the development and defense of plants, such as tissue pigmentation, prevention against damage from UV radiation and mainly against mitigation of oxidative stress. bearing in mind that recent studies using tomatoes (*Solanum lycopersicum*) have shown that transcriptional regulators of key enzyme genes play critical roles in the biosynthesis of flavonoids in tomatoes, such as the overexpression of *SlMYB14*, which increased the expression levels of several structural genes in the phenylpropanoid pathway, modulating the accumulation of flavonoids, in particular resulted in increased levels of flavonols and anthocyanins (Li et al., 2016). Therefore, this *SlMYB14* overexpression improved the tolerance of transgenic plants to 2,4,6-trichlorophenol (2,4,6-TCP), an organic pollutant that can cause serious oxidative damage to plants, making it possible to state that flavonoids can play a role in maintaining ROS homeostasis at the plant (Li & Yao, 2021).

Although we have highlighted the role of compounds from secondary plant metabolism and enzymes responsible for detoxification, these defense mechanisms work concurrently with other antioxidant protection tools, as shown by the study by El-Esawi et al., (2020), where through the inoculation of soybean plants with the microorganism *Serratia marcescens*

BM1 promotes the potential for phytoremediation of Cd by soybean, as well as stress tolerance. In this research, it is highlighted that the exposure to the metal reduced the growth of the plant, as well as biomass, absorption of nutrients, antioxidant capacity and the content of chlorophyll, total phenolics, flavonoids among other parameters. Cd also induced oxidative stress, assessed through the activity and expression of antioxidant enzymes (CAT, APX, SOD and POD), ROS levels and lipid peroxidation. All of these parameters had their values optimized after the inoculation of soybean plants stressed with Cd with *Serratia marcescens BM1*. Including the significant increase in the activities of antioxidant enzymes, osmolyte levels and the expression of stress-related genes in plants with oxidative stress and the levels of total phenolics and flavonoids themselves.

In this perspective, the results found in the literature allow us to affirm that although plants have different xenobiotic detoxification strategies and consequently in neutralizing the initiation and propagation of oxidative stress, the evaluation of these mechanisms for different xenobiotics and plants is a valid parameter for the selection of plants for phytoremediation criteria.

6. Influence of Light on Defense Mechanisms

Changes in light quality provide different biochemical responses that can vary between different species of plants. These responses are of great importance in recent cultivation techniques since the feasibility of adjusting the lighting spectra allows controlling the growth, development and nutritional quality of the plant (Lin et al., 2013).

The development of the plant is strongly influenced by the quality of the light, which refers to the color or the wavelength that is irradiated. It has been shown that the use of blue and red-light emitting diodes (LED) light has been efficient in increasing the fresh and dry weight of cotton (*Gossypium hirsutum* L.), while separately, the red LED induces root growth and blue LED is important for the chlorophyll synthesis and seedling stoma development (Li et al., 2010). Later, another study demonstrated that the addition of yellow light to the treatments with combined spectra of red and blue LEDs promoted the growth of potato seedlings *in vitro*, with the red LED separately, having the most advantageous results for the growth of this plant (Li et al., 2018).

The red and blue light stand out in the positive influence of these parameters, as demonstrated by Kim et al., (2004), where there was an elongation in the stem of chrysanthemum buds with stimulation through a red LED. On the other hand, there are indications that adverse effects may also be caused by the incidence of red LED light alone, as in the growth of the stem and elongation of the shoots in different plants (Hahn et al., 2000; Heo et al., 2002). This can be attributed to the imbalance in the distribution of light energy available for Photosystems I and II, thereby inhibiting plant growth (Tennessen et al., 1994). As well as in the study by Heo et al. (2002), who through the evaluation of the effects generated in marigold and sage generated by blue, red or mixed monochromatic radiation from a fluorescent lamp with light emitting diodes (LEDs) (blue, red or red distant) noticed that the dry weight of marigold seedlings was significantly increased in monochromatic red light, fluorescent light plus red LED or fluorescent light, but reduced with blue monochromatic light. Meanwhile, in sage the dry weight was significantly higher under fluorescent light plus blue LED, fluorescent light plus red LED and fluorescent light plus distant red LED. Thus, it is possible to assume that the effects caused on plant growth by the incidence of wavelengths related to specific or combined colors may have different behaviors depending on the plant species and the combination of light stimulus.

Regarding the influence on the secondary metabolism of higher plants, the quality of light has a pronounced effect on the accumulation of secondary metabolites, however, the mechanisms of such effects are not fully understood. However, the total content of phenolics and carotenoids are among the main effects on secondary plant metabolism, as research developed to date indicates that the quality of light significantly affects the production of these molecules (Alrifai et al., 2019).

From what appears in the literature, apparently, the use of red and blue light are the main colors capable of altering the production of these compounds of the secondary metabolism, considering that changes in the content of the secondary metabolism found by Nadeem et al., (2019) through *Ocimum basilicum* callus treatments with blue, red, green, yellow and white LED. The authors found that the best levels of total phenolics were observed in the callus culture grown under blue light compared to the control, while the maximum accumulation of total flavonoid content was observed in the callus culture grown under red light, as well as the highest amount of anthocyanins (peonidin and cyanidin). While the specific effects of enzyme activities modulated by LED light are limited (Alrifai et al., 2019).

Changes in the total content of phenolics and carotenoids are among the main effects on secondary plant metabolism, as research developed to date indicates that the quality of light significantly affects the production of molecules with strong antioxidant activity (Alrifai et al., 2019). In this sense, studies show that the total content of phenolics and flavonoids has been increased with the treatment of blue (Kapoor et al., 2018) and red (Khurshid et al., 2020) LEDs. Stimulating the production of these compounds directly potentiates the plant's antioxidant action, as presented by Qian et al., (2016) through treatments with different LED colors with *Brassica oleracea var. alboglabra* Bailey, noticed that the treatment with blue LED increased the levels of carotenoids and flavonoids more, and consequently the antioxidant potential.

In addition to the influence of light on the variation in the composition of metabolites, light is an essential factor for plant health, as it also changes the activity of antioxidant defense enzymes, as in the activity of SOD, CAT, G-POD in *Anoectochilus roxburghii* and *Camptotheca acuminata* (Ye et al., 2017; Yu et al., 2017) and the enzymes APX, DHAR, MDHAR and GR in *Eleutherococcus senticosus* (Shohael et al., 2006) which are the most efficient defense mechanisms against free radicals in plants . In this perspective, the use of light stimuli can influence physiological mechanisms, stimulating the growth of the stem, root, in addition to increasing the biomass and the weight of the plants (dry and fresh). Along with this, light is still able to stimulate the production of secondary metabolism compounds, which added to the mentioned antioxidant protection enzymatic mechanisms are capable of gauging the plant's greater resistance. All of these parameters are important when we evaluate the resistance of different plants to the toxicity of xenobiotics, leading us to hypothesize that the application of light stimuli could positively influence the phytoremediation action of several plant species.

Among the light treatments, we highlight the LED although light sources, such as fluorescent and incandescent lamps are often used for growing plants to increase the levels of photosynthetic photon flow, they have wavelengths located outside the spectrum used in photosynthesis, and in this way of low quality to promote benefits for the plants (Kim et al., 2004). A more efficient light source for the construction of lighting systems are the Light Emitting Diodes (LED), which have several advantages compared to the light sources used for the cultivation of plants, such as small mass and volume, durability, long useful life operational, specificity and narrow wavelength bands, low heat generation and photon output that is linear with the incoming electric current. These light sources, therefore, are ideal for use in plant lighting projects and allow wavelengths to be combined with the plant's photoreceptors (Lin et al., 2013).

7. Influence of Light on Phytoremediation

As previously mentioned, light plays an important role in the metabolism of higher plants, stimulating several defense mechanisms that modulate the stress arising from xenobiotic intoxication during phytoremediation, attributing an improvement to the plant's phytoremediation action. Thus, the use of LEDs to stimulate phytoremediation is a promising technique, however there are no records in the literature that evaluate this combination in organic compounds, such as pesticides. On the other hand, although it is scarce, there are some studies in the literature that assess the influence of LED on the phytoremediation of metals.

Luo et al., (2019) evaluated the influence of the combination of red and blue light using five combinations of blue and red light (0%, 10%,

25%, 50% and 100% blue) on the ability of Eucalyptus globulus to concomitantly with different heavy metals. The variations in the combinations of blue and red light resulted in different responses in the plant. The plant's biomass increased as the highest proportion of blue light was used, on the other hand, red light alone provided the highest root / aerial part ratio possibly through greater storage of nutrients (organic fructose and acids), increasing stabilization of metals in the plant's rhizosphere and also biomass in the roots. This is confirmed by the greater accumulations of Cu and Cd in the roots of plants treated with only red LED. Kwon et al., (2017) also found similar results with the use of a red LED in the phytoremediation of metals by microalgae *Phaeodactylum tricornutum, Nitzschia sp. Skeletonema sp.* and *Chlorella vulgaris*, where the latter obtained greater migration and absorption of Cu and Zn under red light, however, blue light provided greater extraction of the metals in question for the other species. This study unfortunately did not provide an explanation of mechanisms that could have led to these results.

On the other hand, the study by Niu et al., (in press) investigated variations in Dissolved Organic Matter (DOM) in the rhizosphere of *Noccaea caerulescens* when irradiated with different combinations of red (0, 25, 50, 90 and 100% red) and blue light, as well as revealing the mechanisms by which light irradiation improves the decontamination effect of Cd by *N. caerulescens*. The study showed that DOM absorption increased as the proportion of red light increased, with DOM fragments in the rhizosphere of *N. caerulescens* it was possible to notice that the hydrophilic portion of the rhizosphere increased in the same way, as the Cd has a greater affinity for the part hydrophilic, the greater amounts of Cd absorbed and stored in the roots of the plants is related to higher proportions of irradiated red light, as well as to higher concentrations of dissolved organic matter, increasing the activity of Cd in the soil.

These results show us that the adequate irradiation of light can improve the phytoremediation efficiency of several plants, increasing their growth rate and absorption capacity, as well as the stimulation of the protection mechanisms already mentioned. Since no study evaluating phytoremediation pesticides with LED stimulus has been carried out to date, we bring up the ROS levels of *P. neochilus* root, stem and leaf after a period of 28 days in contact with the herbicide 2,4-D, in order to assess the toxicity generated by the herbicide in the plant, and whether treatments with different colors of light can contribute to the reduction of oxidative stress caused by 2,4-D. Through this, we can corroborate with the literature records and apply the influence of different lengths of lights to assist in the phytoremediation of pesticides, such as 2,4-D.

8. Effects of LEDs on Reactive Oxygen Species (ROS) Levels

We can see that in Figure 7A-1 there was a generation of ROS levels in the leaves, in all treatments with 2,4-D, being indifferent to the projected light, when compared to natural light. Increased production and accumulation of ROS can damage the cellular components of the plant, such as DNA, lipids, proteins and sugars (Asada, 2006; Baskar et al., 2018; Van Breusegem & Dat, 2006). On the other hand, when we observed only the ROS variations of the treatment with 2,4-D in relation to the control, the red-light treatment (Figure 7A-2) obtained a significant reduction in ROS generation when compared to the other light treatments. That is, its ROS reduction levels are equivalent to those of natural light. Thus suggesting that the emission of red light has a photoprotective effect due to the possible transport of flavonoids inside and outside the cell, which are more located on the wall and vacuole of epidermal cells, which are more dedicated to filtering the most energetic solar wavelengths in reaching the leaf (Agati et al., 2013). Because it is known that many flavonoids are effective in protecting the plant against the absorption of stressors because of absorption in the wavelength and increase the concentrations in the epidermal layers, promoting biosynthesis, which can be regulated by sunlight (Hernández et al., 2009).

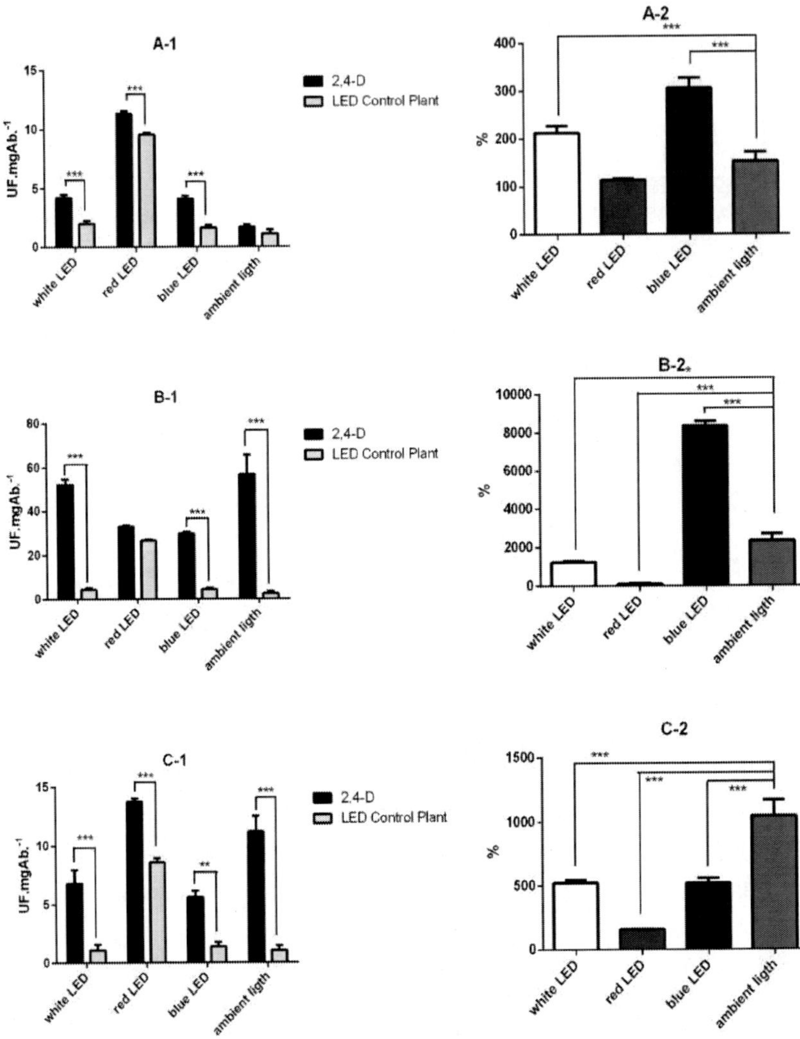

Figure 7. Levels of reactive oxygen species (ROS) in the leaves (A-1), roots (B-1) and stems (C-1) of *P. neochilus* conditioned to the environment with 2,4-D and photostimulation in white, blue light and red and ambient light and their respective controls, under the same lighting conditions, without the addition of pesticides in the water. A-2, B-2 and C-2 present the variation in the ROS levels of the treatments with their respective controls. The lines indicate the comparisons with P < (0.05) (*), P < 0.01 (**) and P < 0.001 (*). The methodology is described in the appendix.

However, when we observed the generation of ROS in the roots, it was found that the emission of red light did not allow the increase in ROS levels (Figure 7B-1), being equivalent to the control, differently from the other light emissions, which raised the ROS levels. In addition, the emission of red light at the root made it possible to further reduce the ROS levels compared to the control of natural light (Figure 7B-2). According to Agati et al., (2013) speculate that flavonol metabolism may have evolved, mainly to prevent and combat excess oxidative stress induced by light, regardless of the different solar wavelengths that reach the earth's surface, as desiccation and high light irradiance were the first challenges faced by plants on earth.

Regarding the stem, we can observe that the increase in ROS occurs in all treatments with 2,4-D, regardless of the type of light (Figure 7C-1) when compared to the control. However, all light, blue, red and white emissions through phytoremediation had a decrease in ROS levels, when compared to natural light (Figure 7C-2), highlighting the red light that further decreases the generation of reactive species in the face of the other light treatments. The results by Lu et al., (2019) go against our hypothesis, where they observed that an increase in the emission of blue light in relation to the emission of red light in branches produced greater biomass and increased the effectiveness of phytoremediation when compared to natural light. In addition, after stress conditions in the plant, the levels of flavonoids are over-stimulated and enabling them to act as an antioxidant in order to balance the plant's completeness (Panche et al., 2016).

Thus, there are mechanisms that help prevent oxidative stress in the plant, using efficient signals to keep ROS under control. Generating secondary metabolites, more specifically phenols, which will adjust these ROS concentrations by activating a network of biochemical events to increase tolerance and thus contributing to their photosynthesis (Chaves et al., 2020). In addition, through what was reported in the present study, we can assume that at the same time that the various secondary metabolism compounds are acting to protect the plant, enzymatic protection mechanisms may be acting concomitantly, since the colors of light influence not only in the phenolic composition (Nadeem et a., 2019). But also, in the activity of antioxidant enzymes modulating oxidative stress, where the results by Shohael et al., (2006) show us that the treatments with red LED light provide a greater activity of GR, GST and SOD, important enzymes responsible for the detoxification of xenobiotics, and in the defense against ROS. Thus, we can say that phytoremediation can indeed be accepted as an economical environmental restoration technology, and that the use of LED light stimuli can contribute to the phytoremediation of pesticides, not just metals.

Conclusion

The present review brought results from the literature that allows us to affirm that plants have protection mechanisms against oxidative stress from xenobiotic intoxication, which may be enzymatic as enzymes responsible for detoxification of ROS or relative to Phases I, II and III referring to metabolic transformations of xenobiotics, or non-enzymatic. non-enzymatic defense mechanisms are products of the secondary metabolism of plants that, through their production, make it possible to modulate the oxidative stress generated by toxic agents. Along with this, light plays an important role in physiological parameters and in the mechanisms mentioned, so that its use during the period of contact of the plant with the toxic agent provides greater resistance to it, and an increase in the phytoremediation capacity of metals. Although there are no studies evaluating the use of lights to stimulate the phytoremediation of organic compounds such as pesticides, our results exposed in this study demonstrate that the red light was able to reduce the oxidative damage in *P. neochilus* caused by 2,4-D possibly through stimulating physiological properties and defense mechanisms, opening an important perspective for the use of lights in stimulating the phytoremediation of organic compounds.

REFERENCES

Abdel-Shafy, H. I., & Mansour, M. S. (2018). Phytoremediation for the elimination of metals, pesticides, PAHs, and other pollutants from wastewater and soil. In Kumar, V., Kumar, M., & Prasad, R. (Eds.), *Phytobiont and ecosystem restitution* (pp. 101-136). Springer, Singapore.

Agati, G., Brunetti, C., Di Ferdinando, M., Ferrini, F., Pollastri, S., & Tattini, M. (2013). Functional roles of flavonoids in photoprotection: new evidence, lessons from the past. *Plant Physiology and Biochemistry*, *72*, 35-45.

Alché, J. D. (2019). A concise appraisal of lipid oxidation and lipoxidation in higher plants. *Redox biology*, *23*, 101136.

Al Mahmud, J., Hasanuzzaman, M., Nahar, K., Bhuyan, M. B., & Fujita, M. (2018). Insights into citric acid-induced cadmium tolerance and phytoremediation in *Brassica juncea* L.: coordinated functions of metal chelation, antioxidant defense and glyoxalase systems. *Ecotoxicology and environmental safety*, *147*, 990-1001.

Alrifai, O., Hao, X., Marcone, M. F., & Tsao, R. (2019). Current review of the modulatory effects of LED lights on photosynthesis of secondary metabolites and future perspectives of microgreen vegetables. *Journal of agricultural and food chemistry*, *67*(22), 6075-6090.

Alscher, R. G., Donahue, J. L., & Cramer, C. L. (1997). Reactive oxygen species and antioxidants: relationships in green cells. *Physiologia Plantarum*, *100*(2), 224-233.

Ansari, A. A., Naeem, M., Gill, S. S., & AlZuaibr, F. M. (2020). Phytoremediation of contaminated waters: An eco-friendly technology based on aquatic macrophytes application. *The Egyptian Journal of Aquatic Research*, *46*(4), 371-376.

Apel, K., & Hirt, H. (2004). Reactive oxygen species: metabolism, oxidative stress, and signal transduction. *Annual Review of Plant Biology*, *55*, 373-399.

Asada, K. (2006). Production and scavenging of reactive oxygen species in chloroplasts and their functions. *Plant physiology*, *141*(2), 391-396.

Aubertheau, E., Stalder, T., Mondamert, L., Ploy, M. C., Dagot, C., & Labanowski, J. (2017). Impact of wastewater treatment plant discharge on the contamination of river biofilms by pharmaceuticals and antibiotic resistance. *Science of the Total Environment*, *579*, 1387-1398.

Bashir, F., Siddiqi, T. O., & Iqbal, M. (2007). The antioxidative response system in *Glycine max* (L.) Merr. exposed to Deltamethrin, a synthetic pyrethroid insecticide. *Environmental pollution*, *147*(1), 94-100.

Baskar, V., Venkatesh, R., & Ramalingam, S. (2018). Flavonoids (antioxidants systems) in higher plants and their response to stresses. In Gupta, D. K., Palma, J. M., & Corpas, F. J. (Eds.), *Antioxidants and antioxidant enzymes in higher plants* (pp. 253-268). Springer, Cham.

Chaves, N., Santiago, A., & Alías, J. C. (2020). Quantification of the antioxidant activity of plant extracts: analysis of sensitivity and hierarchization based on the method used. *Antioxidants*, *9*(1), 76.

Chrétien, F., Giroux, I., Thériault, G., Gagnon, P., & Corriveau, J. (2017). Surface runoff and subsurface tile drain losses of neonicotinoids and companion herbicides at edge-of-field. *Environmental pollution*, *224*, 255-264.

Coleman, J., Blake-Kalff, M., & Davies, E. (1997). Detoxification of xenobiotics by plants: chemical modification and vacuolar compartmentation. *Trends in plant science*, *2*(4), 144-151.

Collins, C., Fryer, M., & Grosso, A. (2006). Plant uptake of non-ionic organic chemicals. *Environmental science & technology*, *40*(1), 45-52.

Daneshvar, A., Aboulfadl, K., Viglino, L., Broséus, R., Sauvé, S., Madoux-Humery, A. S., & Prévost, M. (2012). Evaluating pharmaceuticals and caffeine as indicators of fecal contamination in drinking water sources of the Greater Montreal region. *Chemosphere*, *88*(1), 131-139.

Deng, Z., & Cao, L. (2017). Fungal endophytes and their interactions with plants in phytoremediation: a review. *Chemosphere*, *168*, 1100-1106.

Dietz, A. C., & Schnoor, J. L. (2001). Advances in phytoremediation. *Environmental health perspectives*, *109*(suppl 1), 163-168.

Dordio, A. V., & Palace Carvalho, A. J. (2011). Phytoremediation: An option for removal of organic xenobiotics from water. In Golubev, I. A. (Ed.). *Handbook of Phytoremediation* (pp. 51-92). Nova Science Publishers, NY, USA.

Duran, R. E., Kilic, S., & Coskun, Y. (2015). Response of maize (*Zea mays* L. *saccharata* Sturt) to different concentration treatments of deltamethrin. *Pesticide biochemistry and physiology*, *124*, 15-20.

Dushenkov, V., Kumar, P. N., Motto, H., & Raskin, I. (1995). Rhizofiltration: the use of plants to remove heavy metals from aqueous streams. *Environmental science & technology*, *29*(5), 1239-1245.

Eapen, S., Singh, S., & D'souza, S. F. (2007). Advances in development of transgenic plants for remediation of xenobiotic pollutants. *Biotechnology Advances*, *25*(5), 442-451.

Ekperusi, A. O., Sikoki, F. D., & Nwachukwu, E. O. (2019). Application of common duckweed (*Lemna minor*) in phytoremediation of chemicals in the environment: State and future perspective. *Chemosphere*, *223*, 285-309.

El-Esawi, M. A., Elkelish, A., Soliman, M., Elansary, H. O., Zaid, A., & Wani, S. H. (2020). *Serratia marcescens* BM1 enhances cadmium stress tolerance and phytoremediation potential of soybean through modulation of osmolytes, leaf gas exchange, antioxidant machinery, and stress-responsive genes expression. *Antioxidants*, *9*(1), 43.

Garg, N., & Manchanda, G. (2009). ROS generation in plants: boon or bane?. *Plant Biosystems*, *143*(1), 81-96.

Gutiérrez-Grijalva, E. P., Picos-Salas, M. A., Leyva-López, N., Criollo-Mendoza, M. S., Vazquez-Olivo, G., & Heredia, J. B. (2018). Flavonoids and phenolic acids from oregano: Occurrence, biological activity and health benefits. *Plants*, *7*(1), 2.

Hahn, E. J., Kozai, T., & Paek, K. Y. (2000). Blue and red light-emitting diodes with or without sucrose and ventilation affect in vitro Growth of *Rehmannia glutinosa* plantlets. *Journal of Plant Biology*, *43*(4), 247-250.

Heo, J., Lee, C., Chakrabarty, D., & Paek, K. (2002). Growth responses of marigold and salvia bedding plants as affected by monochromic or mixture radiation provided by a light-emitting diode (LED). *Plant Growth Regulation*, *38*(3), 225-230.

Hernández, I., Alegre, L., Van Breusegem, F., & Munné-Bosch, S. (2009). How relevant are flavonoids as antioxidants in plants?. *Trends in plant science*, *14*(3), 125-132.

Hussain, F., Hadi, F., & Rongliang, Q. (2021). Effects of zinc oxide nanoparticles on antioxidants, chlorophyll contents, and proline in *Persicaria hydropiper* L. and its potential for Pb phytoremediation. *Environmental Science and Pollution Research*, 1-17.

Igamberdiev, A. U., & Lea, P. J. (2002). The role of peroxisomes in the integration of metabolism and evolutionary diversity of photosynthetic organisms. *Phytochemistry*, *60*(7), 651-674.

Jeevanantham, S., Saravanan, A., Hemavathy, R. V., Kumar, P. S., Yaashikaa, P. R., & Yuvaraj, D. (2019). Removal of toxic pollutants from water environment by phytoremediation: a survey on application and future prospects. *Environmental technology & innovation*, *13*, 264-276.

Kapoor, S., Raghuvanshi, R., Bhardwaj, P., Sood, H., Saxena, S., & Chaurasia, O. P. (2018). Influence of light quality on growth, secondary metabolites production and antioxidant activity in callus culture of *Rhodiola imbricata* Edgew. *Journal of Photochemistry and Photobiology B: Biology*, *183*, 258-265.

Kapsi, M., Tsoutsi, C., Paschalidou, A., & Albanis, T. (2019). Environmental monitoring and risk assessment of pesticide residues in surface waters of the Louros River (NW Greece). *Science of the Total Environment*, *650*, 2188-2198.

Kasote, D. M., Katyare, S. S., Hegde, M. V., & Bae, H. (2015). Significance of antioxidant potential of plants and its relevance to therapeutic applications. *International journal of biological sciences*, *11*(8), 982.

Khurshid, R., Ullah, M. A., Tungmunnithum, D., Drouet, S., Shah, M., Zaeem, A., & Abbasi, B. H. (2020). Lights triggered differential accumulation of antioxidant and antidiabetic secondary metabolites in callus culture of *Eclipta alba* L. *Plos one*, *15*(6), e0233963.

Kidd, K. A., Blanchfield, P. J., Mills, K. H., Palace, V. P., Evans, R. E., Lazorchak, J. M., & Flick, R. W. (2007). Collapse of a fish population after exposure to a synthetic estrogen. *Proceedings of the National Academy of Sciences*, *104*(21), 8897-8901.

Kidd, K. A., Paterson, M. J., Rennie, M. D., Podemski, C. L., Findlay, D. L., Blanchfield, P. J., & Liber, K. (2014). Direct and indirect responses of a freshwater food web to a potent synthetic oestrogen. *Philosophical Transactions of the Royal Society B: Biological Sciences*, *369*(1656), 20130578.

Kim, S. J., Hahn, E. J., Heo, J. W., & Paek, K. Y. (2004). Effects of LEDs on net photosynthetic rate, growth and leaf stomata of chrysanthemum plantlets in vitro. *Scientia Horticulturae*, *101*(1-2), 143-151.

Korte, F., Kvesitadze, G., Ugrekhelidze, D., Gordeziani, M., Khatisashvili, G., Buadze, O., & Coulston, F. (2000). Organic toxicants and plants. *Ecotoxicology and environmental safety*, *47*(1), 1-26.

Kovacs, H., & Szemmelveisz, K. (2017). Disposal options for polluted plants grown on heavy metal contaminated brownfield lands–A review. *Chemosphere*, *166*, 8-20.

Kvesitadze, G., Khatisashvili, G., Sadunishvili, T., & Ramsden, J. J. (2006). *Biochemical mechanisms of detoxification in higher plants: basis of phytoremediation*. Springer Science & Business Media.

Kwon, H. K., Jeon, J. Y., & Oh, S. J. (2017). Potential for heavy metal (copper and zinc) removal from contaminated marine sediments using microalgae and light emitting diodes. *Ocean Science Journal*, *52*(1), 57-66.

Li, H., Xu, Z., & Tang, C. (2010). Effect of light-emitting diodes on growth and morphogenesis of upland cotton (*Gossypium hirsutum* L.) plantlets in vitro. *Plant Cell, Tissue and Organ Culture (PCTOC)*, *103*(2), 155-163.

Li, R., Huang, W., Wang, X., Liu, X., & Xu, Z. (2018). Effects of yellow, green, and different blue spectra on growth of potato plantlets in vitro. *HortScience*, *53*(4), 541-546.

Li, Z., Peng, R., & Yao, Q. (2021). *SlMYB14* promotes flavonoids accumulation and confers higher tolerance to 2, 4, 6-trichlorophenol in tomato. *Plant Science*, *303*, 110796.

Lin, K. H., Huang, M. Y., Huang, W. D., Hsu, M. H., Yang, Z. W., & Yang, C. M. (2013). The effects of red, blue, and white light-emitting diodes on the growth, development, and edible quality of hydroponically grown lettuce (*Lactuca sativa* L. var. *capitata*). *Scientia Horticulturae*, *150*, 86-91.

Liu, N., Zhong, G., Zhou, J., Liu, Y., Pang, Y., Cai, H., & Wu, Z. (2019). Separate and combined effects of glyphosate and copper on growth and antioxidative enzymes in *Salvinia natans* (L.) All. *Science of the Total Environment*, *655*, 1448-1456.

Luo, J., Cao, M., Zhang, C., Wu, J., & Gu, X. S. (2020). The influence of light combination on the physicochemical characteristics and enzymatic activity of soil with multi-metal pollution in phytoremediation. *Journal of hazardous materials*, *393*, 122406.

Luo, J., He, W., Wu, J., & Gu, X. S. (2019). Sensitivity of *Eucalyptus globulus* to red and blue light with different combinations and their influence on its efficacy for contaminated soil phytoremediation. *Journal of environmental management*, *241*, 235-242.

Meyer, J. M., Rezende, F. M., Soares, S. A., & Tomba, A. (2013). Metabolismo Secundário [Secondary Metabolism]. In Lopez, A. M. (Org.). *Botânica no Inverno* [*Botany in winter*] (pp.34-41), São Paulo.

Mierziak, J., Kostyn, K., & Kulma, A. (2014). Flavonoids as important molecules of plant interactions with the environment. *Molecules*, *19*(10), 16240-16265.

Mittler, R. (2002). Oxidative stress, antioxidants and stress tolerance. *Trends in plant science*, *7*(9), 405-410.

Mitton, F. M., Gonzalez, M., Monserrat, J. M., & Miglioranza, K. S. (2018). DDTs-induced antioxidant responses in plants and their influence on phytoremediation process. *Ecotoxicology and environmental safety*, *147*, 151-156.

Mitton, F. M., Miglioranza, K. S., Gonzalez, M., Shimabukuro, V. M., & Monserrat, J. M. (2014). Assessment of tolerance and efficiency of crop species in the phytoremediation of DDT polluted soils. *Ecological engineering*, *71*, 501-508.

Nadeem, M., Abbasi, B. H., Younas, M., Ahmad, W., Zahir, A., & Hano, C. (2019). LED-enhanced biosynthesis of biologically active ingredients in callus cultures of *Ocimum basilicum*. *Journal of Photochemistry and Photobiology B: Biology*, *190*, 172-178.

Nahler, G. (2009). International Conference on Harmonisation (ICH). In *Dictionary of Pharmaceutical Medicine* (pp. 96-96). Springer, Vienna.

Niu, Y., Wang, Z., Wang, H., Yang, X., Cao, M., & Luo, J. [in press]. Influence of light-irradiated *Noccaea caerulescens* on the characteristics of dissolved organic matter in its rhizospheric soil during phytoremediation. *Research Square*.

Noctor, G., Mhamdi, A., & Foyer, C. H. (2014). The roles of reactive oxygen metabolism in drought: not so cut and dried. *Plant physiology*, *164*(4), 1636-1648.

Panche, A. N., Diwan, A. D., & Chandra, S. R. (2016). Flavonoids: an overview. *Journal of nutritional science*, *5*.

Patra, D. K., Pradhan, C., & Patra, H. K. (2019). Chromium bioaccumulation, oxidative stress metabolism and oil content in lemon grass *Cymbopogon flexuosus* (Nees ex Steud.) W. Watson grown in chromium rich over burden soil of Sukinda chromite mine, India. *Chemosphere*, *218*, 1082-1088.

Patra, D. K., Pradhan, C., & Patra, H. K. (2020). Toxic metal decontamination by phytoremediation approach: concept, challenges, opportunities and future perspectives. *Environmental Technology & Innovation*, *18*, 100672.

Pereira, R. J., & Cardoso, M. D. G. (2012). Vegetable secondary metabolites and antioxidants benefits. *J. Biotechnol. Biodivers*, *3*(4), 146-152.

Ramborger, B. P., Gularte, C. A. O., Rodrigues, D. T., Gayer, M. C., Carriço, M. R. S., Bianchini, M. C., ... & Roehrs, R. (2017). The phytoremediation potential of *Plectranthus neochilus* on 2, 4-dichlorophenoxyacetic acid and the role of antioxidant capacity in herbicide tolerance. *Chemosphere*, *188*, 231-240.

Ramírez-Sandoval, M., Melchor-Partida, G. N., Muñiz-Hernández, S., Girón-Pérez, M. I., Rojas-García, A. E., Medina-Díaz, I. M., ... & Velázquez-Fernández, J. B. (2011). Phytoremediatory effect and growth of two species of *Ocimum* in endosulfan polluted soil. *Journal of hazardous materials*, *192*(1), 388-392.

Rmiki, N.-E.; Lemoine, Y.; Schoefs, B. (1999) Carotenoids and Stress in Higher Plants and Algae. In Pessarakli, M. (Eds.), *Handbook of plant and crop stress* (Ed. 2. pp 465-482). New York: Marcel Dekker.

Rosa, A. D. S. (2013). *Fitorremediação de pesticidas utilizados em lavouras de arroz através do cultivo hidropônico de alface (Lactuca sativa L.)* [*Phytoremediation of pesticides used in rice crops using culture hydroponic lettuce (Lactuca sativa L.)*] [Master's thesis, Federal University of Pampa]. Institutional Repository of Unipampa.

Sandermann Jr, H. (1994). Higher plant metabolism of xenobiotics: the 'green liver' concept. *Pharmacogenetics*, *4*(5), 225-241.

Santos, J. S., da Silva Pontes, M., Grillo, R., Fiorucci, A. R., de Arruda, G. J., & Santiago, E. F. (2020). Physiological mechanisms and phytoremediation potential of the macrophyte *Salvinia biloba* towards a commercial formulation and an analytical standard of glyphosate. *Chemosphere*, *259*, 127417.

Santos, L. H., Araújo, A. N., Fachini, A., Pena, A., Delerue-Matos, C., & Montenegro, M. C. B. S. M. (2010). Ecotoxicological aspects related to the presence of pharmaceuticals in the aquatic environment. *Journal of hazardous materials*, *175*(1-3), 45-95.

Saxena, P., Hiwrale, I., Das, S., Shukla, V., Tyagi, L., Pal, S., & Dhodapkar, R. (2021). Profiling of emerging contaminants and antibiotic resistance in sewage treatment plants: An Indian perspective. *Journal of Hazardous Materials*, *408*, 124877.

Schneider, J. R., Müller, M., Klein, V. A., Rossato-Grando, L. G., Barcelos, R. P., Dalmago, G. A., & Chavarria, G. (2020). Soybean Plant Metabolism under Water Deficit and Xenobiotic and Antioxidant Agent Application. *Biology*, *9*(9), 266.

Shohael, A. M., Ali, M. B., Yu, K. W., Hahn, E. J., Islam, R., & Paek, K. Y. (2006). Effect of light on oxidative stress, secondary metabolites and induction of antioxidant enzymes in *Eleutherococcus senticosus* somatic embryos in bioreactor. *Process Biochemistry*, *41*(5), 1179-1185.

Silva, M. L. C., Costa, R. S., dos Santos Santana, A., & Koblitz, M. G. B. (2010). Compostos fenólicos, carotenóides e atividade antioxidante em produtos vegetais [Phenolic compounds, carotenoids and antioxidant activity in plant products]. *Semina: Ciências Agrárias*, *31*(3), 669-681.

Sjerps, R. M., Kooij, P. J., van Loon, A., & Van Wezel, A. P. (2019). Occurrence of pesticides in Dutch drinking water sources. *Chemosphere*, *235*, 510-518.

Song, B., Xu, P., Chen, M., Tang, W., Zeng, G., Gong, J., & Ye, S. (2019). Using nanomaterials to facilitate the phytoremediation of contaminated soil. *Critical reviews in environmental science and technology*, *49*(9), 791-824.

Styszko, K., Proctor, K., Castrignanò, E., & Kasprzyk-Hordern, B. (2021). Occurrence of pharmaceutical residues, personal care products, lifestyle chemicals, illicit drugs and metabolites in wastewater and receiving surface waters of Krakow agglomeration in South Poland. *Science of the Total Environment*, *768*, 144360.

Subedi, B., Balakrishna, K., Joshua, D. I., & Kannan, K. (2017). Mass loading and removal of pharmaceuticals and personal care products including psychoactives, antihypertensives, and antibiotics in two sewage treatment plants in southern India. *Chemosphere*, *167*, 429-437.

Sultana, T., Murray, C., Kleywegt, S., & Metcalfe, C. D. (2018). Neonicotinoid pesticides in drinking water in agricultural regions of southern Ontario, Canada. *Chemosphere*, *202*, 506-513.

Tangahu, B. V., Sheikh Abdullah, S. R., Basri, H., Idris, M., Anuar, N., & Mukhlisin, M. (2011). A review on heavy metals (As, Pb, and Hg) uptake by plants through phytoremediation. *International Journal of Chemical Engineering*, *2011*.

Tennessen, D. J., Singsaas, E. L., & Sharkey, T. D. (1994). Light-emitting diodes as a light source for photosynthesis research. *Photosynthesis research*, *39*(1), 85-92.

Vafaei, F., Movafeghi, A., Khataee, A. R., Zarei, M., & Lisar, S. S. (2013). Potential of *Hydrocotyle vulgaris* for phytoremediation of a textile dye: inducing antioxidant response in roots and leaves. *Ecotoxicology and environmental safety*, *93*, 128-134.

Van Breusegem, F., & Dat, J. F. (2006). Reactive oxygen species in plant cell death. *Plant physiology*, *141*(2), 384-390.

Vasconcelos, P. S. P. M. (2011). *Compostos fenólicos e potencial antioxidante de ervas consumidas na região amazônica brasileira* [*Phenolic compounds and antioxidant potential of herbs consumed in the Brazilian Amazonian region*]. [Master's thesis, Campinas State University] Unicamp's Scientific and Intellectual Production Repository.

Ververidis, F., Trantas, E., Douglas, C., Vollmer, G., Kretzschmar, G., & Panopoulos, N. (2007). Biotechnology of flavonoids and other phenylpropanoid-derived natural products. Part II: Reconstruction of multienzyme pathways in plants and microbes. *Biotechnology Journal: Healthcare Nutrition Technology*, *2*(10), 1235-1249.

Wang, J., Chen, X., Chi, Y., Chu, S., Hayat, K., Zhi, Y., & Zhou, P. (2020). Optimization of NPK fertilization combined with phytoremediation of cadmium contaminated soil by orthogonal experiment. *Ecotoxicology and environmental safety*, *189*, 109997.

Wang, Q., Ma, L., Zhou, Q., Chen, B., Zhang, X., Wu, Y., & Feng, Y. (2019). Inoculation of plant growth promoting bacteria from hyperaccumulator facilitated non-host root development and provided promising agents for elevated phytoremediation efficiency. *Chemosphere*, *234*, 769-776.

Winkel-Shirley, B. (2001). It takes a garden. How work on diverse plant species has contributed to an understanding of flavonoid metabolism. *Plant physiology*, *127*(4), 1399-1404.

Wrubleswski, J., Reichert, F. W., Galon, L., Hartmann, P. A., & Hartmann, M. T. (2018). Acute and chronic toxicity of pesticides on tadpoles of *Physalaemus cuvieri* (Anura, Leptodactylidae). *Ecotoxicology*, *27*(3), 360-368.

Yang, X., Feng, Y., He, Z., & Stoffella, P. J. (2005). Molecular mechanisms of heavy metal hyperaccumulation and phytoremediation. *Journal of trace elements in medicine and biology*, *18*(4), 339-353.

Yang, Y., Ok, Y. S., Kim, K. H., Kwon, E. E., & Tsang, Y. F. (2017). Occurrences and removal of pharmaceuticals and personal care products (PPCPs) in drinking water and water/sewage treatment plants: A review. *Science of the Total Environment*, *596*, 303-320.

Ye, S., Shao, Q., Xu, M., Li, S., Wu, M., Tan, X., & Su, L. (2017). Effects of light quality on morphology, enzyme activities, and bioactive compound contents in *Anoectochilus roxburghii*. *Frontiers in Plant Science*, *8*, 857.

You, J., & Chan, Z. (2015). ROS regulation during abiotic stress responses in crop plants. *Frontiers in plant science*, *6*, 1092.

Yu, W., Liu, Y., Song, L., Jacobs, D. F., Du, X., Ying, Y., & Wu, J. (2017). Effect of differential light quality on morphology, photosynthesis, and antioxidant enzyme activity in *Camptotheca acuminata* seedlings. *Journal of Plant Growth Regulation*, *36*(1), 148-160.

Zhu, Y. X., Du, W. X., Fang, X. Z., Zhang, L. L., & Jin, C. W. (2020). Knockdown of *BTS* may provide a new strategy to improve cadmium-phytoremediation efficiency by improving iron status in plants. *Journal of hazardous materials*, *384*, 121473.

APPENDIX – METHODOLOGY

2,4-D Treatments

The treatments were carried out by placing 2 seedlings of *Plectranthus neochilus* in glass flasks with 500 ml of water (with or without herbicide), each flask consists of a replicate of each group. Each treatment consisted of a group of plants with 2,4-D (commercial formulation Aminol 806 at a concentration of 1,206 $kg.he^{-1}$), containing two seedlings of *P. neochilus* and water with 2,4-D and the group of control of the plant to, containing only plants without the presence of pesticide. All groups will be made in triplicate. The plants were collected 28 days after the beginning of the exposure with the herbicide and divided into leaves, stem and root and kept at -80ºC for future analysis.

Levels of Reactive Oxygen Species (ROS)

To carry out the biochemical tests, 0.5 grams of root, stem or leaf of the different treatments (individually) were homogenized with 1 mL of Tris-HCl 50 mmol.L^{-1}, the homogenate was centrifuged at 11000 RPM at 4°C and the supernatant was collected and frozen at -80°C for future analysis. The levels of reactive oxygen species were also assessed using the protocol described by Shohael et al., (2006) with our adaptations. Initially 20 µL of sample were mixed with 160 µL of water and 20 µL of 2,7-dichlorofluorecein (DCF) 1 mmol.L^{-1} prepared moments before use in dimethylsulfoxide (DMSO). The reading was performed for 30 minutes every 15 seconds by fluorescence with an emission wavelength of 525 nm and excitation of 485 nm, after which 20 µL of H_2O_2 was added and the reading was again performed.

Statistical Analysis

To compare the ROS measurement data of the 2,4-D treatments with their controls, two-way ANOVA with Bonferroni's post test was used. To assess the difference in ROS variations by the DCF assay, one-way ANOVA with Tukey's post hoc was used.

In: The Future of Phytoremediation
Editor: James C. Flores
ISBN: 978-1-53619-625-2

Chapter 3

PISTIA STRATIOTES AS AN OPTION FOR AQUATIC PHYTOREMEDIATION

***Bruna Piaia Ramborger*[1]*, Maria Elizabeth Gomes Paz*[1]*, Ketelin Monique Cavalheiro Kieling*[1]*, Elton Luis Gasparotto Denardin*[2] *and Rafael Roehrs*[1]**

[1]Laboratory of Chemical, Environmental and Toxicological Analysis (LAQAT), Federal University of Pampa, Uruguaiana - RS -Brazil
[2]Laboratory of Physico-Chemical Studies and Natural Products (LEFQPN), Federal University of Pampa, Uruguaiana - RS -Brazil

ABSTRACT

Aquatic contamination is a worldwide problem resulting from organic and inorganic compounds of agricultural, domestic, or industrial origin. The types of contaminants found in the aquatic environment vary between heavy metals, pesticides, medicines, industrial residues, among others. Among the forms of decontamination, phytoremediation is the most convenient, due to its practicality, low costs, and being ecologically accepted. This technique uses plant metabolism to remove, detoxify, or immobilize environmental contaminants. Some plants are highly resistant to toxicity and can absorb high concentrations of different compounds from the environment and therefore stand out in the efficiency of phytoremediation. *Pistia stratiotes* has these peculiarities, and for this reason, over the years, researchers have used it to promote phytoremediation of the most diverse contaminants. This plant is a free-floating aquatic macrophyte, from tropical and subtropical ecosystems. It can be an invasive aquatic species, as it can completely cover the watercourses forming dense mats. Although this characteristic makes it inconvenient, several studies have shown that it is precisely for this reason that it can be used for phytoremediation and subsequent production of biofuel. Thus, studies with this plant date back to 1962 and are still being carried out. In this chapter, we will present data related to the anatomical characteristics, popular use, chemical composition, and forms of growth of *P. stratiotes*, in addition to phytoremediation processes. The toxicological and metabolic parameters related to phytoremediation with this plant will also be addressed, as well as the different studies that used *P. stratiotes* for environmental remediation in waters contaminated with heavy metals and organic compounds.

Keywords: wastewater, decontamination, biomass, metals, pesticides, biogas

1. INTRODUCTION

Pistia stratiotes, popularly known as water lettuce, is a floating aquatic monocot macrophyte free of freshwater. There are indications that its origin is South America (Renner and Zhang 2004) since it is free-floating, has modified reproduction structures, and has a wide geographical distribution. It develops areas of the tropical and subtropical climate mainly, having an easy and fast reproduction in these conditions. Due to these peculiarities (high reproduction, growth, and free-floating in slow rivers and reservoirs), their increased growth can generate a dense layer of biomass, causing problems for animals and plants native to the body of water (Brunel et al., 2010).

Although this plant is classified by the European and Mediterranean Organization Plant Protection (EPPO) as a plant with a high impact on the environment and priority (pest control sector) (Brunel et al., 2010; EPPO, 2017), several studies are using *P. stratiotes* for other purposes (Table 1).

In Brazil, Ricardo et al., (2017) and Silva Filho and Brandão (1992) described this plant used in folk medicine for abscess treatment, skin diseases, and diuretic purposes. In addition to its use in popular medicine, there is research on the variety of bioactive compounds present in this plant (Daroon and Haroon, 2012; Sinang et al., 2019; Tyagi and Agarwal, 2017) and on the production of biofuels such as biogas, biohydrogen, and ethanol from fermentation of plant biomass by microorganisms (Nipaney and Panholzer, 1987; Abbasi et al., 1991; Mishima et al., 2008; Anuradha, Abbasi and Abbasi, 2015; Pantawong et al., 2015; Yu-Chen, Zheng and Cheng, 2015; Gusain and Suthar, 2017; Mthethwa et al., 2018; Mthethwa et al., 2019). The most highlight of the use of *P. stratiotes* is related to its effectiveness in phytoremediation, as this technique has a beneficial impact on the environment through the treatment of urban, rural, or domestic effluents (Olguín et al., 2017; Sridhar, 1986; De Campos et al., 2019; Lin and Li, 2016; Ribeiro et al., 2019). Table 1 presents these studies with *P. stratiotes* in brief.

2. DESCRIPTION OF *P. STRATIOTES*

Pistia stratiotes (Araceae), known as water lettuce, is a free-floating aquatic macrophyte from tropical and subtropical ecosystems. It is native to the southern region of the American continent and has spread rapidly to other areas. It is usually found in still water and in places that receive a lot of solar radiation. This plant can reproduce through sexual reproduction, or, mainly, by clonal reproduction, by breaking the connections between branches (Cardoso et al., 2005). In determining locations, *P. stratiotes* is an invasive aquatic species because it can completely cover watercourses. In this way, this plant can damage irrigation systems, hinder navigation, and damage hydroelectric plants (Cilliers et al., 1996; Winton and Clayton, 1996; Cancian et al., 2009; Cardoso et al., 2005; Hill and Coetzee, 2017; Rezania et al., 2017). Figure 1 shows the anatomy of the root and leaf of *P. stratiotes* and its vegetative cycle during the seasons. According to Rosa et al., (2019), the plant root is fasciculate, and the stem is of the stolon type, and both stay submerged in the water. The leaf has an average size of 5 cm, is incomplete, only with spongy limbus its phyllotaxis is rosy and with actinodroma venation.

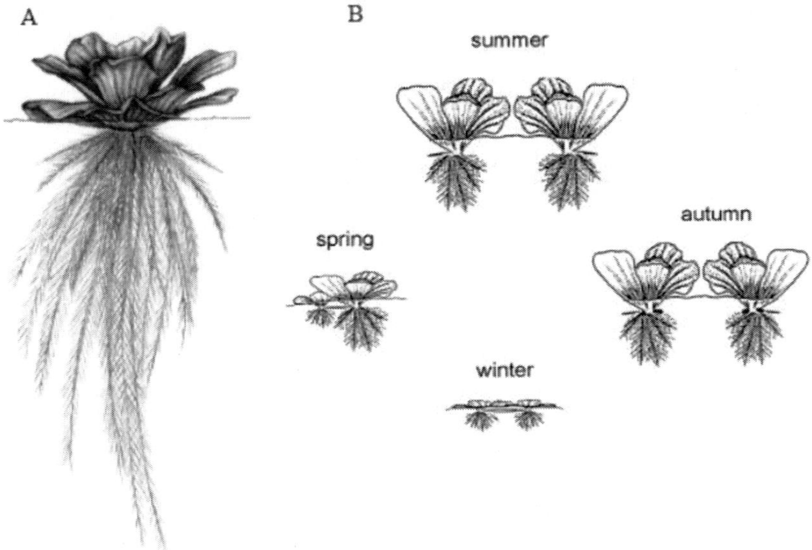

Source: A - Tavares (2002), B - Hussner, Heidbuechel, and Heiligtag (2014).

Figure 1. Anatomy of the root and leaf of *Pistia stratiotes* and their life cycle in the Erft River (North Rhine-Westphalia, Germany).

According to Neuenschwander et al., (2009), this plant consists of a rosette of leaves obovate to spatulate, velvety, light green, and covered with short hair. The leaves retain air bubbles that allow them to float (Figure 1 A and B). The underside of the leaves is densely hairy and almost white, with longitudinal ribs. A clonal plant can form small colonies through stolons. Inflorescences are imperceptible (7 - 12.5 mm), have short peduncles in the center of the rosette, which grow on a stem. The spadix, closed in a whitish spathe, is light green, hairy on the outside, and glabrous on the inside. The spathe usually shows a constriction between the groups of male and female flowers. When *P. stratiotes* is in a very toxic environment, it can show visual symptoms, such as chlorosis and root zone reduction (Figure 2).

Table 1. Studies with *P. stratiotes* over the years

Subject matter	Period	References*
Chemical control over *P. stratiotes*	1962 - 1991	Service (1962); Pablico and Moody (1991); Cilliers et al., (1991)
Forms of cultivation, growth, and propagation in environments	1966 - 2014	Chadwick and Obeid (1966); Attionu (1976); Silva (1981); Nelson, Smith and Best (1981); Sharma and Sridhar (1981); Sridhar and Sharma (1985); Dray Jr and Center (1989); Agami and Reddy (1990); Dewald and Lounibos (1990); Harley (1990); Sajna et al. (2007); Cancian, Camargo and Silva, 2009; Hussner, Heidbuechel and Heiligtag (2014)
Composition and biochemistry	1977 - 2019	Zennie and Mcclure (1977); Greca et al., (1991); Monaco and Previtera (1991); Nakata, Kostman and Franceschi (2003); Daroon and Haroon (2012); Jha, Sharma, and Ganesh (2013); Tyagi and Agarwal (2017); Sinang et al., (2019)
Popular use and research	1980 - 2017	Toursarkissian, 1980; Silva Filho and Brandão, 1992; Achola, Indalo and Munenge, 1997; Lahitte et al., 1998; Mukhtar and Tukur, 2001; Mukhtar and Huda, 2003; Kumar et al.,2010; Pallavi et al., 2011; Kumar et al., 2011; Daboor and Haroon, 2012; Koffuor et al., 2012; Kyei et al., 2012; Jha, Sharma, and Ganesh (2013); Ricardo et al., (2017)
Phytoremediation	1985 - 2020	De et al., (1985), Sridhar (1986); Satyakala and Jamil (1992); Odjegba and Fasidi (2004); Miretzky, Saralegui and Fernandez (2004); Gujarathi, Haney and Linden (2006); Zimmels, Kirzhner and Malkovskaja (2006); Mishra and Tripathi (2008); Tewari et al. (2008); Espinoza-Quinõnes et al., (2009); Lu et al., (2010); Akinbile and Yusoff (2011); Das, Goswami and Talukdar (2013); Pérez et al., (2013); Farnese et al. (2013); Aurangzeb et al., 2014; Farnese et al. (2014); Hanks, Caruso and Zhang (2015); Putra, Cahyana and Novarita (2015); Di Luca et al., (2016); Lin and Li (2016); Olguín et al., (2017); Sudiarto, Renggamen and Choi (2019); Wang et al., (2018); De Campos et al., (2019); Escoto et al., (2019); Kumar et al., (2019); Novita, Moersidik and Priadi (2019); Ribeiro et al., (2019); Tabinda et al., (2020).

Table 1. (Continued)

Subject matter	Period	References*
Production of biogas, ethanol, fermentable sugar, biohydrogen, nanoparticles synthesis and biocombustível	1987 - 2019	Nipaney and Panholzer (1987); Abbasi, Nipaney and Panholzer (1991); Mishima et al. (2008); Anuradha, Abbasi and Abbasi (2015); Pantawong et al., (2015); Yu-Chen, Zheng and Cheng (2015); Gusain and Suthar (2017); Mthethwa et al., (2018); Mthethwa et al., (2019).
Morphological and biological changes	2000 - 2013	Tarlyn et al., (1998); Keates et al. (2000); Kostman, Franceschi and Nakata (2003); Sinha, Saxena and Singh (2005); Upadhyay and Panda (2005); Tewari et al. (2008); Upadhyay and Panda (2009); Farnese et al., (2013)
Characterization and morphology	2007 - 2019	Cicero et al. (2007); Neuenschwander et al., (2009); Silva et al., (2013); Rosa et al., (2019)

* The authors' quotation is in chronological order.

Source: De Campos et al., (2019).

Figure 2. Visual symptoms of As (III) toxicity in *Pistia stratiotes*: chlorosis, darkening, and reduction of the root system. (A) leaves, (B) roots. 1 - Control, 2 - 5 μM As (III), 3 - 10 μM As (III), 4 - 20 μM As (III).

2.1. Popular Use and Research

Despite the unwanted environmental impacts that *Pistia stratiotes* can cause, it is an advantageous alternative when used to treat illnesses. In traditional medicine, the use of its leaves as an emollient and for ophthalmic purposes (Toursarkissian, 1980), inflammations (especially hemorrhoids) (Lahitte et al., 1998), and the complete plant for diuretic purposes (Silva Filho and Brandão, 1992). Its leaves are used to treat abscesses and other skin diseases in Brazil (Ricardo et al., 2017). Research related to the use of its extracts in animal models has proven its application for analgesic, antipyretic and anti-inflammatory purposes (Koffuor et al., 2012; Kumar et al., 2011, Kyei et al., 2012). As well as for wound healing

(Jha et al., 2013), as a diuretic (Pallavi et al., 2011), anthelmintic (Kumar et al., 2010) and bronchodilator (Achola et al., 1997). *In vitro* analyzes showed that *P. stratiotes* is useful in stomach disorders treatment, inflammation of the throat and mouth, and exert antimicrobial and antifungal action (Daboor and Haroon, 2012; Mukhtar and Tukur, 2000; Mukhtar and Huda, 2003).

2.2. Known Phytochemical Composition

Research that carried out the phytochemical screening of *Pistia stratiotes* revealed several bioactive compounds that enhance its medicinal use. Some of these compounds were: alkaloids, anthocyanins, tannins, phenols, terpenoids, glycosides, cardiotonic glycosides, quinones, anthraquinones, resins, sterols, and volatile compounds (Daroon and Haroon, 2012; Sinang et al., 2019; Tyagi and Agarwal, 2017). Besides, chromatographic analyzes (GC-MS) of *P. stratiotes* leaves identified the following compounds: n-hexadecanoic acid, hexadecanoic acid, ethyl ester, 9,12,15-octadecatrienoic acid, methyl ester, hexadecanoic acid, 2- ester hydroxy-1-(hydroxymethyl) ethyl, diisooctyl phthalate, stigmasterol, and L-glutamine (Tyagi and Agarwal, 2017). All of these findings show the pharmacological importance of this plant.

2.3. Easy Growing Conditions (Laboratory and Field Scale)

Many experiments with *P. stratiotes* are carried out on a laboratory scale and generally use nutrient solutions to improve development and growth. This solution uses different concentrations of the following nutrients: NH_4; NO_3; PO_4; K; Here; Mg; Faith; H_3BO_3; $MnCl_2$ $4H_2O$; $ZnSO_4$ 7H2O; $CuSO_4$ $5H_2O$; H_2Mo_2 $4H_2O$ according to Hoagland and Arnon (1950). Regarding the pH of the medium, Chadwick and Obeid (1966) showed that the highest production of the dry mass of *P. stratiotes* occurred at pH 4, although the species still survives at higher pH values.

However, doing so shows a considerable reduction in vigor. The pH value changes according to the location, the organic matter, and microbiological activity. In phytoremediation experiments, the *P. stratiotes* generally make the pH neutral after a few days (Akinbile and Yusoff, 2011; Novita, Moersidik and Priadi, 2019). When *P. stratiotes* is in the environment, the better growth response is when the plant is at 25°C and a 12 hour photoperiod (Cancian et al., 2009).

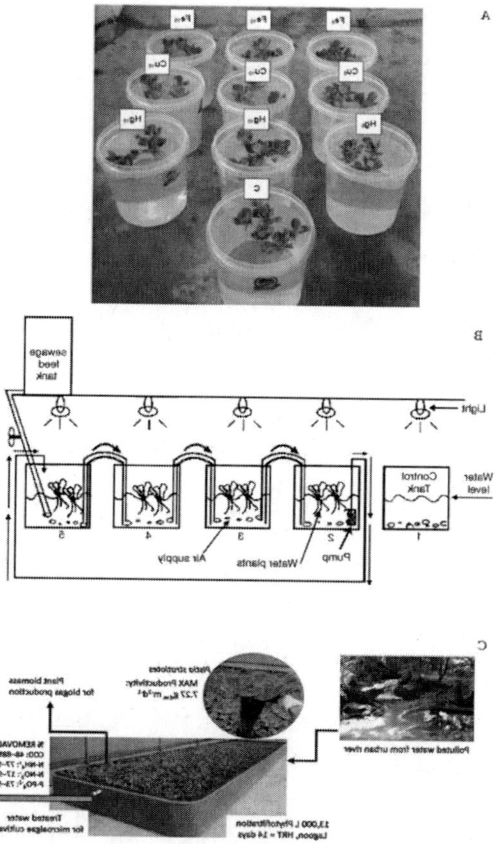

Figure 3. Forms used to conduct experiments with *P. stratiotes*. A - laboratory scale, separate tanks (Kumar et al., 2019). B - laboratory scale, circulation-scale tanks (Zimmels et al., 2006). C - Field-scale, phytofiltration pond to supply treated water and plant biomass for biofuels production (Olguín et al., 2017).

In phytoremediation experiments developed on a laboratory scale, *P. stratiotes* grow and reproduce in tanks with nutrient solutions (or not), together with the contaminant. These tanks can be separated for each group to be analyzed or connected like a circuit. Figure 3 (A and B) shows these types of experiments.

When studies are performed on a field scale (*in situ*), *P. stratiotes* is introduced into the contaminated environment (Lu et al., 2010), or phytofiltration ponds are built on site (Figure 3 C) (Olguín et al., 2017). In these types of experiments, *P. stratiotes* is used for phytoremediation processes, and its biomass can be used for other purposes, making the technique even more advantageous.

2.4 Benefits of Unconventional Use - Biogas, Biohydrogen and Ethanol Production

P. stratiotes reproduce very rapidly, thus, producing much biomass. This biomass has relatively high levels of carbohydrates, proteins, fats, and crude fiber. Therefore, some researchers decided to invest in the use of *P. stratiotes* as a renewable raw material. This investment allowed to decrease the problems caused by water resources (Gusain and Suthar, 2017; Kaur et al., 2018).

In this sense, *P. stratiotes* was the object of study as a substrate for biogas production by fermentation reactors. This peculiarity based on the raw material that makes up the plant, which provided a high biogas content, with an average methane content of 58 - 68% (Abbasi et al., 1991; Nipaney and Panholzer, 1987) and 66.35% (Pantawong et al., 2015). The works of Mthethwa et al., (2018) and Mthethwa et al., (2019) showed that the acid hydrolysis of *Pistia stratiotes* with mixed microbial culture promoted the production of biohydrogen through a dark fermentation process. The low cost presented in the study of Mthethwa et al., (2018) was the most interesting (0.08 $/kg dry weight of *P. stratiotes*). In the study by Mishima et al., (2008), a fermentation of *P. stratiotes* was carried out with the microorganisms *Saccharomyces cerevisiae* NBRC 2346 and

Escherichia coli KO11. Ethanol production with these bacterial strains was 14.9 g/L of *S. cerevisiae* and 16.2 g/L of *E. coli*. Also, the ethanol yield per unit of biomass was comparable to those reported for other agricultural biomasses.

3. Phytoremediation

The largest number of studies with *P. stratiotes* remain concentrated in phytoremediation. This technique consists of plant uses, which in association with microorganisms, promote the environment decontamination (soil, water, and air). It is a green technology, as it uses the natural processes of the plant and its associated microbiota to promote the degradation, accumulation, and/or stabilization of organic and inorganic contaminants (De Campos et al., 2019; Gomes, 2012; Rajkumar et al., 2012; Ramborger et al., 2017; Shah and Daverey, 2020; Wei et al., 2021). In this sense, phytoremediation can decontaminate heavy metals, pharmaceutical products, industrial, domestic, or agricultural waste. Therefore, it is an operationally simple, economically viable, and aesthetically pleasing strategy. What makes phytoremediation even more beneficial is its implementation, which does not require specialized personnel or expensive state-of-the-art equipment (Shah and Daverey, 2020). And it is for these reasons that the use of *P. stratiotes* is so advantageous.

3.1. Phytoremediation Processes

Phytoremediation is classified according to the plant metabolic process or by these processes associated with the root bacteria. The types of phytoremediation are briefly explained below and illustrated in Figure 4.

- *Phytoextraction:* refers to the accumulation of soil contaminants within the tissues of the plant. In this case, the plant can absorb and translocate the contaminant into its tissues. It can be called phytoaccumulation, phyto-absorption, or phytosequestration. (Ali et al., 2013; Tripathi et al., 2020).

- *Rhizofiltration:* occurs in an aquatic environment and is similar to phytofiltration. In this way, the plants' location is where they absorb contaminants and concentrate them on the roots and shoots (Kumar et al., 2019).
- *Phytodegradation:* it is the accumulation and degradation of organic contaminants through the metabolic activity of the plant. It can be accomplished by the enzyme's dehalogenase and oxygenase, for example (Ali et al., 2013).
- *Rhizodegradation:* refers to organic pollutant's decontamination in the soil by microorganisms in the rhizosphere. The rhizosphere is the part of the plant root with the surrounding soil surface and is where microorganisms increase plant growth directly or indirectly. Therefore, they are symbiotic relationships in which the plant roots provide exudates containing carbohydrates, amino acids, and flavonoids that serve as nutrients (carbon and nitrogen sources) to stimulate microbial activity, which will promote the degradation of pollutants. Besides, plants also release certain enzymes in the rhizosphere that can degrade organic contaminants in soils (Ali et al., 2013; Kumar et al., 2019; Tripathi et al., 2020).
- *Phytovolatilization:* is the absorption phenomenon of pollutants and their release into the atmosphere in a volatile form. This technique applies to organic compounds and some heavy metals like Hg and Se. However, the phytovolatilization only transfers the contaminant from one matrix (soil or water) to another (atmosphere). Therefore, it does not remove the compound totally, and it can be redeposited (Ali et al., 2013; Wiszniewska et al., 2016).
- *Phytostabilization:* promotes the trapping of contaminants in a matrix (soil or water) through adsorption (Wiszniewska et al., 2016).

- *Phyto-stimulation/Rhizostimulation:* occurs when native plant species are explored, or sown new species, appropriate to the type of soil and climate. Thus, it is possible to promote the stimulation of microorganisms in this soil. However, it requires inorganic nutrients, organic co-substrates, compaction, among others, to promote remediation (Kumar et al., 2019).
- *Phytodesalinization:* refers to the use of halophyte plants to remove salts from soils affected by salts, to enable them to support plant growth (Ali et al., 2013; Wiszniewska et al., 2016).

Figure 4 summarizes the forms of phytoremediation that can be performed in water.

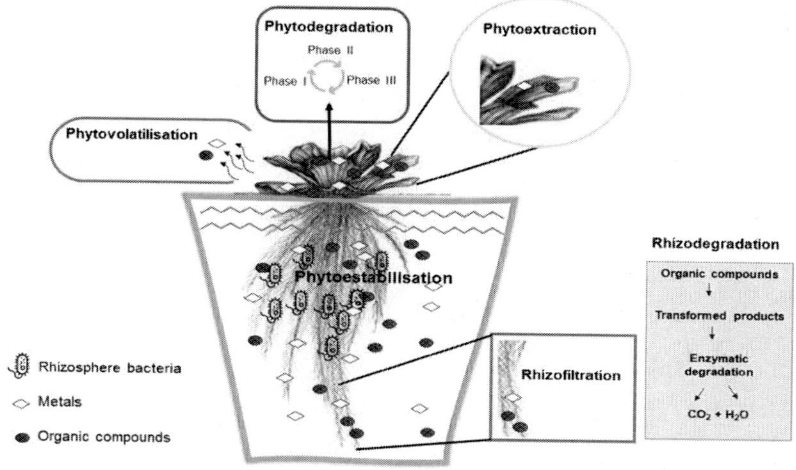

Source: Authors, adapted from Kumar et al. (2018).

Figure 4. Phytoremediation processes.

As can be seen, phytoremediation processes are due to the natural plant's performance in their environment. In this sense, most of the pollutants surrounding the plant are absorbed by the roots and are translocated to the leaves (Wang and Liu, 2007). Unless that intensive spraying occurs for the absorption occurs directly through the leaves

(Burken et al., 2005). Generally, contaminants access the plant's internal system via diffusion and are led by the xylem flow. The transport rate within the plant is influenced by the physical and chemical properties of the contaminant (volatility, miscibility in water, electrical charge, density, and size) and by interactions with the matrix system because the plant's internal systems are devoid of specific carriers (Fujisawa, 2002). Therefore, some precautions are necessary for the application of phytoremediation on a large scale. Some examples are a detailed characterization of the site, the selection of specific plant species for phytoremediation at the target site, the cost-benefit analysis (growth, planting, and seedling management), and the estimated time for remediation. All of these issues are the subject of studies in research with different plants and contaminants. These details are present in several studies that used *P. stratiotes* to remove organic and inorganic pollutants (section 4).

4. Phytoremediation with *Pistia stratiotes:* Toxicological and Metabolic Parameters

According to Table 1 (introduction section), most studies conducted with *P. stratiotes* address its ability to absorb contaminants from the environment (phytoremediation). Over the years, many studies have evaluated the phytoremediation potential of *P. stratiotes* with other responses produced by it, such as toxicity and biochemicals response. The next topics and Table 2 will briefly address these surveys.

4.1. Phytoremediation of Heavy Metals

Contamination of the aquatic environment by heavy metals is a problem for the world, as these compounds are highly toxic to plants and animals. Heavy metals, unlike organic pollutants, are persistent, so they tend to accumulate in the environment. Its source may be in the mining, in urban sewage, in foundries, in tanneries, the textile industry, and the chemical industry. Different techniques can remove them from environments, such as reverse osmosis, ion exchange, chemical precipitation, adsorption, and solvent extraction. However, these practices are generally not environmentally friendly, and their operating and maintenance costs are high (Ali et al., 2020). In this sense, the green technology employed by phytoremediation becomes a viable alternative because it is efficient, aesthetically pleasing, economical, and environmentally friendly (Kumari et al., 2015). Therefore, several studies have verified whether the use of *P. stratiotes* to remove different heavy metals from aquatic environments would be successful.

One of the first studies that used phytoremediation with *P. stratiotes* was by Anil et al., (1985), who verified the removal of 90% Hg (II) in concentrations below 20 ppm. Shortly afterward, Sridhar's (1986) work investigated the plant's ability to accumulate heavy metals in Lake Awba (Nigeria). Since this lake contained a variety of waste (laboratory, laundry, cafeteria, kitchen, and student hostels), this study showed that *P. stratiotes* was able to remove and store significant amounts of cadmium, chromium, cobalt, copper, lead, nickel, and manganese from the aquatic environment. Sridhar (1986) intended to use the plant for animal feed after this phytoremediation. However, this was not allowed due to the high concentrations of these metals in the plant. On a laboratory scale, Satyakala and Jamil (1992) observed a more accumulation of chromium (Cr) in the roots, followed by leaves with an increase in Cr concentrations (10, 20, and 50 ppm) in a 72-hour exposure period. And Miretzky et al., (2004) reported that *P. stratiotes* removed 97.3% of Cr of 1 mg/L, 72.2% of 2 mg/L, and 73.5% of 4 mg/L of metal solutions prepared in 190h.

After a period, research results improved, and the tests with *P. stratiotes* were carried out with more amount of metal or in an association of them. In this sense, Odjegba and Fasid (2004) found that in low concentrations, *P. stratiotes* removed Ag, Cd, Cr, Cu, Hg, Ni, Pb and, Zn from the water in 21 days. Also, it was observed the macrophyte efficiency with the removal of 96%, 95%, 92%, 82%, and 81% of Cu 1 mg/L, Fe 2 mg/L, Zn 2 mg/L, Cd 2 mg/L, and Cr 1 mg/L (Mishra and Tripathi, 2008). Espinoza et al., (2009) quantified lead (Pb) in the aquatic (hydroponic) environment with *P. stratiotes* and, they obtained results from modeling the bioaccumulation kinetics of the metal in the living plant. These results served for the work of Pérez et al., (2013), who used microcosms in semi-continuous flow to evaluate the remediation of P. stratiotes on a laboratory scale. The synthetic effluent used obtained remediation of 76.8 and 81.3% for Pb (II); 81.1 and 69.9% for Cr (III). When in the association, the phytoremediation of Pb (II) was 73.4% for Pb (II) and 77.9% for Cr (III). After phytoremediation, the final Cr (III) concentration was below the maximum limit allowed by the regulation for discharge into natural water bodies (2 mg/L). For Pb (II), just the initial concentration of 1 mg/L Pb complies with what was established by that legislation (0.5 mg/L). Das et al., (2013) observed a maximum extraction of Cadmium (Cd) by the plant of up to 11.67 mg/g dry weight of the root and 5.75 mg/g dry weight of the leaves in the exposure doses of 10 and 15 mg/L, respectively. Putra et al., (2015) combined an electro-assisted system with hydroponic phytoremediation (*P. stratiotes*) and obtained the removal of approximately 99% of Pb and Cu in 7 days. In this experiment, the electrode in water contaminated with Pb and Cu allowed the metal migration from the anode to the cathode. Then, this promotes more mobilization of metals in the root system of *P. stratiotes*. Aurangzeb et al., (2014) showed that *P. stratiotes* had a greater affinity for Pb and Cu in a steel effluent, with removal rates of 70.7% and 66.5%, respectively. And De Campos et al., (2019) suggested that the plant may have a potential for phytoremediation of aquatic environments contaminated with arsenic (As III) at a concentration of 10 μM. Besides, the success of this macrophyte was in phytoremediation of Cu (II) (53.20%), Fe (III) (83.20%), and Hg (II) (62.14%), at 5 mg/L (Kumar et al., 2019). Finally, Tabinda et al., (2020) showed different removal rates for Cu and Cr: 91.20% and 77.30%, respectively.

According to the studies described, *P. stratiotes* experiments are carried out on a laboratory scale or in a contaminated environment (*in situ*). The experiments carried out on a laboratory scale are extremely important because it is possible to evidence the potential of the plant's phytoremediation, the toxicity, and the biochemical effects triggered by the contaminants. Thus, through these studies, the implementation of *P. stratiotes* in the contaminated site (*in situ*), as industrial waste and domestic sewage, becomes viable. Concerning surveys carried out *in situ*, in addition to showing the removal of contaminants present in these locations, they also showed an improvement in water quality with the phytoremediation process (Aurangzeb et al., 2014; Di Luca et al., 2016; Mishra and Tripathi, 2008; Pérez et al., 2013). The physical-chemical parameters evaluated in these cases are pH, dissolved oxygen levels (DO), biochemical oxygen demand (BOD), total alkalinity, total dissolved solids, and electrical conductivity. A pH and total alkalinity concentration decreased concerning the control with Pb (II) 1 mg/L (Pérez et al., 2013). It can be related to organic acid production from the cycle of growth, death, and organic matter mineralization. In Pb (II) 5 mg/L treatment, there was a decrease in pH and DO index. Regarding the treatment of Cr (III) 4 and 6 mg/L, higher pH values, total alkalinity, and DO were observed concerning the control group. Besides, there were no observed changes in the appearance of the plants. In Aurangzeb et al., (2014), *P. stratiotes* decreased the physical-chemical parameters of a steel foundry effluent (Fazenda Industrial Hayatabad, Peshawar) as follows: pH: effluent 9.1; *P. stratiotes* 8.5; total dissolved solids: effluent 430; *P. stratiotes* 356; electrical conductivity: effluent 900; *P. stratiotes* 714.

Other phytoremediation studies with *P. stratiotes* are the toxicity of metals to the plant, such as the survival rate, visual symptoms (color and size of roots and leaves), and the biochemical process involved as enzymes, proteins, and chlorophylls. In the case of Anil et al., (1985), at concentrations of 0.05, 1, and 2 ppm of Hg (II), *P. stratiotes* survived for seven days, but in the highest concentration used (20 ppm), it survived only two days. The toxicity of Hg 20 ppm was through the following analyzes: decrease in chlorophyll (a, b, and total), protein (aerial part), free amino acids (roots), RNA, dry weight, and activities of catalase, protease, and alkaline pyrophosphatase. There was an increase in free amino acids (shoot), peroxidase activity, and acid pyrophosphatase. These results were concerning the control group that refers to the plant that was not in contact with Hg. At concentrations of Hg (II) below 20 ppm, these constituents were less affected, thus showing this plant's ability for environmental decontamination of this heavy metal. Satyakala and Jamil (1992) obtained chlorosis and leaf stripping of *P. stratiotes* for phytoremediation of Cr at concentrations of 25 and 50 ppm. The chlorophyll, sugar, and protein contents decrease with a Cr concentration increase. Also, there was an increase in the phenol content of the plant. This increase was justified as toxicity by Cr since the accumulation of the metal interrupted cell membranes, causing injury and an increase in the concentration of several phenolic compounds (Fry, 1979). Enzymatic activity, succinate dehydrogenase, glutathione peroxidase, and polyphenol oxidase showed a higher cellular respiration rate. The superoxide dismutase and glutathione reductase decreased in the leaves and increased in the roots. In this sense, Satyakala and Jamil (1992) showed the bioaccumulation of high concentrations of Cr in plant tissues. However, this concentration led to the displacement or replacement of metal ions. This accumulation resulted in changes in photosynthetic and respiratory activities. Tewari et al., (2008) made it possible to reduce metals from municipal sludge by *P. stratiotes.* The plant combated the damage generated by metals through the antioxidant enzyme synthesis and other metal binders. This result showed the suitability of the plant for phytoremediation residues.

Odjegba and Fasidi (2004) showed that root elongation and the emergence of new roots, hitherto not observed in previous studies, showed a significant decrease according to the type of metal, concentration, and exposure time. Briefly, the increasing order of root growth at the concentration of 0.3 mM was: Hg, Ni, Cu, Cd, Pb, Ag, Cr e, Zn. Additionally, new roots appeared only with P. stratiotes from the control group. The study authors explained that this reduction could have occurred due to the high solute potential of the external environment, which may prevent the plant from absorbing nutrients, direct inhibition of enzymes of physiological importance, or mitotic division inhibition of meristematic cells. Regarding morphology, both leaf area and biomass production were lower in P. stratiotes treated with metals. Although this effect is typical for all metals in the study, the most prominent was Hg. Besides, there was an inverse relationship between the relative growth rate of the plant with metal concentrations. This fact also was observed in previous studies with heavy metals. Farnese et al., (2014) carried out a complete toxicity and biochemical parameters study of *P. stratiotes* with water contaminated by arsenic (As). In this way, *P. stratiotes* was in contact with As in the concentrations of 0, 5, 10, 15, and 20 μM for seven days. In general, the results showed a sharp drop in the growth rate of the plants and a decrease in the content of photosynthetic pigments (chlorophyll A, total chlorophyll, and carotenoids). As described by the authors, this decrease in the growth rate is a typical plant response to exposure to As in high concentrations. This response can also be related to increased lipid peroxidation, and biochemical results showed the toxicity. The lowest arsenic concentration increased superoxide dismutase, peroxidase (antioxidants), and amino acid contents. In the highest concentrations, there was a decrease. Also, morphological changes occurred in the aerial and the roots (concentrations of 15 and 20 μM). There were chlorosis and leaf necrosis, secondary roots formation decreased, root loss, and the root system darkening. Therefore, this study could show that *P. stratiotes* promoted phytoremediation of As, although this compound presents toxicity in the plant.

In the Hanks et al., (2015) work, *P. stratiotes* was grown in media enriched with different concentrations of silver nanoparticles (AgNP) and silver ions (Ag I) (0.02, 0.2, and 2 mg/L). The results showed that *P. stratiotes* could survive healthily in solutions with AgNP and Ag only in the lowest concentration (0.02 mg/L). In the other concentrations, there was acute toxicity in the plant leaves. However, the plant phytoabsorbed the contaminants. The toxicity occurred due to the absence of nutrients in the aqueous medium. Therefore, these plants may have experienced additional stress. If nutrients were available, they would help maintain the plant's normal internal metabolic processes or combat the Ag destructive. Besides, Hanks et al., (2015) suggested that, on a large scale, this concentration of Ag would be divided among several plants and would reduce its toxic implications, allowing an increase in phytoremediation.

Table 2. Studies on phytoremediation with *P. stratiotes*

Phytoremediation (contaminant)	Biochemical and other effects (plant)	Evaluated symptoms (plant)	Physicochemical parameters evaluated (water)	References
Heavy metals				
Hg (II)	Chlorophyll (a, b, and total), protein (aerial part), free amino acids (roots), RNA, dry weight, and activities of the following enzymes: catalase, protease and alkaline pyrophosphatase	Survival (days x concentration)		De et al., (1985)
Cd, Cr, Co, Cu, Pb, Ni and, Mn			pH and biochemical oxygen demand	Sridhar (1986)
Cr	Chlorophyll, sugar, protein content and phenol content. Enzymatic activity: succinate dehydrogenase, glutathione peroxidase, polyphenol oxidase, superoxide dismutase and glutathione reductase	Chlorosis and leaves shedding		Satyakala e Jamil (1992)
Ag, Cd, Cr, Cu, Hg, Ni, Pb and, Zn		Root elongation and the emergence of new roots. Leaves area and biomass production		Odjegba and Fasidi (2004)
Fe, Cu, Zn, Cr and, Cd			pH, temperature, total dissolved solids, and electrical conductance	Mishra and Tripathi (2008)

Phytoremediation (contaminant)	**Biochemical and other effects (plant)**	**Evaluated symptoms (plant)**	**Physicochemical parameters evaluated (water)**	**References**
Pb, Ni, Cu, Cr, Mn and, Zn	Síntese induzida de enzimas antioxidantes e outros ligantes de metais			Tewari et al., (2008)
Pb		Growth rate, wet biomass	pH	Espinoza-Quinõnes et al., (2009)
Pb (II), Cr (III) and, a mixture of Pb (II) with Cr (III)			pH, total alkalinity concentration. Dissolved oxygen	Pérez et al., (2013)
Cd	Bioconcentration factor, translocation factor and translocation efficiency	Survival, leaves color and dry biomass		Das, Goswami and Talukdar (2014)
Pb and Cu			pH, total dissolved solids, electrical conductivity	Aurangzeb et al., (2014)
As	Superoxide dismutase and peroxidase activity. Amino acid, nitrogen and, starch content. Photosynthetic pigment content (chlorophyll A, total chlorophyll, and carotenoids), mesophyll thickness,	Plant growth rate, formation of secondary roots and darkening of the root system		Farnese et al., (2014)

Table 2. (Continued)

Phytoremediation (contaminant)	Biochemical and other effects (plant)	Evaluated symptoms (plant)	Physicochemical parameters evaluated (water)	References
AgNP and Ag (I)		Survival, leaves color, growth rate, stage of a plant's physical degradation and biomass production		Hanks, Caruso, and Zhang, (2015)
Cr	Bioconcentration factor, translocation factor, total Kjeldahl nitrogen and Total phosphorus (TP) Chlorophyll a concentration	Survival and growth rate		Di Luca et al., (2016)
As (III)	Bioconcentration factor, biomass production and, oxigen-reactive species	Chlorosis, darkening and reduction of the root system		De Campos et al., (2019)
Cu (II), Fe (III) and, Hg (II)	Chlorophyll content,	Kinetic growth rate, root size		Kumar et al., (2019)
Cu and Cr		Dry biomass, phytomorphological characteristics		Tabinda et al., (2020)
Organic compounds				
Sewage			Biochemical oxygen demand (BOD), chemical oxygen demand (COD), total suspended solids (TSS) and turbidity	Zimmel, Kirzhn and Malkovska(2006)

Phytoremediation (contaminant)	Biochemical and other effects (plant)	Evaluated symptoms (plant)	Physicochemical parameters evaluated (water)	References
Rainwater in water retention systems			pH, electrical conductivity, turbidity, suspended solids and nutrients (N and P).	Lu et al., (2010)
Carbamazepine, ibuprofen, sulfadiazine, sulfamethoxazol, sulfamethasone, and triclosan		Survival, growth rate and root system	pH	Lin and Li (2016)
Urban river polluted with domestic wastewater	Biomass production		Chemical oxygen demand, ammonium, nitrate, and phosphate content	Olguín et al., (2017)
Pig wastewater			Total nitrogen	Sudiarto, Renggamen and Choi (2019)
Clomazone herbicide (Gamit)		Survival, color and structure of leaves and roots		Escoto et al., (2019)
Hexazinone herbicide	Adaxial epidermis, abaxial epidermal, palisade parenchyma, aerenchyma and leaf blade			Ribeiro et al., (2019)

The lack of nutrients was verified by Di Luca et al., (2016) when observing that the addition of nutrients P and N significantly favored the removal of Cr from the water of a lake and improved the translocation of Cr to the leaves of the plant.

De Campos et al., (2019) showed that *P. stratiotes* could accumulate As III in the roots and leaves at 10 μM. The plant showed a better bioconcentration factor (BCF) when exposed to 5 and 10 μM As III. Also, it obtained good biomass production at the lowest concentration. As evidenced in previous studies with other metals, De Campos et al., (2019) also observed toxicity in the highest concentration of As (III) (20 μM). The results showed visual symptoms of chlorosis, darkening, and reduction of the root system, in addition to biochemical changes (increased levels of membrane damage and reactive oxygen species - ROS). The antioxidant system action (mainly in the leaves) and the compartmentalization of the pollutant in the root system allowed the plants to maintain cellular homeostasis and growth in the two lowest As III concentrations. And again, the data suggested that *P. stratiotes* may have a potential for phytoremediation of aquatic environments contaminated with As III concentrations close to 10 μM. In the study by Kumar et al., (2019), the translocation factor data increased gradually from 8 to 24 days and remained the same in up to 32 days. The concentration of 5 mg/L of each metal showed fresh biomass from the Cu II plant: 45.57; Fe III: 90.82 and, Hg II: 47.27 g; chlorophyll content: Cu II: 2.99 ± 0.03; Fe III: 4.26 ± 0.06 and Hg II: 3.02 ± 0.03 mg/g fwt; kinetic growth rate (Cu II: 0.0004; Fe III: 0.0011 and Hg II: 0.0006 g/g/day; root size: Cu II: 12.28; Fe III: 16.54 and Hg II: 12.86 cm. In general, this study suggested that the remediation more efficient for these metals, and the maximum growth of *P. stratiotes,* can be achieved with 5 mg/L.

A very current data in these studies with heavy metals and *P. stratiotes* is the accumulation of these compounds that occur mainly in the roots and little in the aerial part (Anil et al., 1985; Farnese et al., 2014; Putra et al., 2015; Odjegba and Fasidi, 2004; Pérez et al., 2013; Satyakala and Jamil, 1992; Tabinda et al., 2020). In the case of Anil et al., (1985), the accumulation of Hg (II) was twice higher in the roots than in the aerial part, in higher concentrations. At lower concentrations, the Hg (II) accumulation in the roots was about four times greater than in the shoot. The possible biochemical explanation for this fact was that Hg (II) formed a protein complex with the SH group (Sigel, 1973), promoting its precipitation in the roots, preventing Hg (II) from passing to the aerial part. This process also was observed in Jarvis et al., (1977) study. In the case of Odjegba and Fasidi (2004), the most accumulated metal was Zn, and the one that least, Ni. This fact could be the result of the presence or absence of a specific plant carrier. De Campos et al., (2019) showed that As (III) maintained mainly at the root (TF always less than 1.0). Tabinda et al., (2020) showed that the average accumulation of Cr in the *P. stratiotes* roots and leaves was 85 mg and 56 mg, and for Cu, it was 96 mg and 70 mg. Through these studies, although *P. stratiotes* present characteristic symptoms of toxicity in high concentrations of heavy metals, it still manages to accumulate high rates of these metals, mainly in its roots. This fact is very crucial for environmental control because, on a larger scale, these compounds are hardly highly concentrated, except for industrial accidents, for example. Moreover, many authors suggest using this macrophyte after phytoremediation due to its high biomass production.

4.2. Phytoremediation of Organic Compounds

The phytoremediation applicability with *P. stratiotes* also is implemented to improve water quality. In the work of Zimmels et al., (2006), *P. stratiotes* was able to decrease the biochemical oxygen demand (BOD), chemical oxygen demand (COD), total suspended solids (TSS), and turbidity to levels allowed for water use for irrigation of tree plantations. Besides, these data were better with the circulation of water in containers. Thus, the technique can be applied on a field-scale as well, both in winter and in the summer. The results by Lu et al., (2010) presented the efficacy and phytoremediation potential of *P. stratiotes* in removing nutrients and in the quality of rainwater in water retention systems. *P. stratiotes* was in the treatment plots of two rainwater retention ponds

(East and West Ponds, Florida) and showed a decrease in water turbidity (more than 60%), suspended solids, and phosphorus and nitrogen concentrations. Olguín et al., (2017) also obtained satisfactory results with water quality (COD removal, ammonium, nitrates, and phosphates) and macrophyte biomass productivity in an urban river polluted with domestic wastewater. Olguín et al., (2017) concluded that the phytofiltration pond with *P. stratiotes* is very viable within the biorefinery for providing biomass all year round and for treating polluted water very effectively. One of the latest studies in which applied *P. stratiotes in situ* was the Sudiarto, Renggamen, and Choi (2019). It determined the removal of nitrogen and phosphorus from swine wastewater treated by aquatic plants, including *P. stratiotes*. Sudiarto, Renggamen, and Choi (2019) showed that the macrophyte had a high tolerance in the effluent and removed 63.15% of the total nitrogen.

The *P. stratiotes* phytoremediation almost generated complete tetracycline and oxytetracycline modifications in six days (Gujarathi et al., 2006). The authors suggested that there was the involvement of metabolites secreted by the plant root in these antibiotics' modifications. In Lin and Li (2016), *P. stratiotes* obtained high rates of removal of pharmaceutical and personal care products (PPCPs) in 13 days. Although the high initial concentration of PPCP caused toxic effects on the plant, it promoted phytoremediation in both concentrations (10 mg/L and 0.8 mg/L). The authors verified the types of degradation of each active ingredient: sulfadiazine, sulfamethoxazole, and sulfamethasone were removed moderately by photodegradation, biodegradation, and absorption. Ibuprofen was removed predominantly by biodegradation or absorption because its removal efficiency was highly dependent on aquatic biomass. Triclosan was removed by photodegradation, absorption, and adsorption by roots and leaves due to its high hydrophobicity and non-ionization. Carbamazepine was almost not removed. Therefore, the authors suggested the importance of adequate biomass allocation and plant residence time for PPCP removal efficiency in wet areas (Lin and Li, 2016).

Phytoremediation strategies for pesticides prevent damage to the entire environment and, *P. stratiotes* stand out in studies in aquatic environments. In the study by Escoto et al., (2019), *P. stratiotes* remedied up to 90% of the herbicide clomazone (Gamit) during the experimental period of 24 days. The interesting thing was the difference in the half-life of this herbicide in the groups used. Thus, in the control group, the half-life was 19.6 days and eight days in treatments. This fact characterized the plant's high potential for clomazone phytoremediation. In Ribeiro et al., (2019) study, *P. stratiotes* was not effective in hexazinone herbicide decontamination. In this case, the amount of light was the determining factor for the control group (without the plants) to obtain an advantage in degradation. The positive point of this research was that hexazinone did not alter the *P. stratiotes* morphology at any concentration tested.

Table 2 presents the phytoremediation studies with *P. stratiotes* and the criteria evaluated in each study briefly.

CONCLUSION

Pistia stratiotes reproduce rapidly, forming dense biomass layers in slow rivers and reservoirs in the whole world. This plant is easy to grow, used in traditional medicine and scientific research, as in the production of biogas and biofuel and, mainly, in phytoremediation. *P. stratiotes* promote phytoremediation of heavy metals, pharmaceuticals, and pesticides. Therefore, the use of *P. stratiotes* for phytoremediation is advantageous, simple, and of extreme ecological importance.

REFERENCES

Abbasi, S. A., Nipaney, P. C., and Panholzer, M. B. 1991. "Biogas production from the aquatic weed Pistia (*Pistia stratiotes*)."

Tecnologia Bioresource. 37: 211-214. Accessed January 05, 2021. doi:10.1016/0960-8524(91)90185-M.

Achola, K. J., Indalo, A. A. and Munenge, R. W. 1997. "Pharmacological activities of Pistia stratiotes." *International Journal Pharmacognosy,* 35: 329-333. Accessed January 05, 2021. doi: 10.1080/09251619708951277.

Agami, M., and Reddy, K. R. 1990. "Competition for space between *Eichhornia crassipes* (Mart.) Solms and *Pistia stratiotes* L. cultured in nutrient-enriched water." *Aquatic Botany*. 38: 195-208. Accessed January 12, 2021. doi:10.1016/0304-3770(90)90005-6.

Akinbile, C. O., and Yusoff, Mohd S. 2011. "Assessing water hyacinth (*Eichhornia crassopes*) and lettuce (*Pistia stratiotes)* effectiveness in aquaculture wastewater treatment." *International Journal of Phytoremediation*. 14: 201-211. Accessed January 08, 2021. doi:10.1080/15226514.2011.587482.

Ali, Hazrat, Khan, Ezzat, and Sajad, M. Anwar. 2013. "Phytoremediation of heavy metals concepts and applications." *Chemosphere* 91: 869-881. Accessed January 11, 2021. doi:10.1016/j.chemosphere.2013.01.075.

Ali, Shafaqat, Abbas, Zohaib, Rizwan, Muhammad, Zaheer, I. Elahi, Yava, Ilkay, Ünay, Aydın, Abdel-DAIM, M. M., Bin-Jumah, May, Hasanuzzaman, Mirza and Kalderis, Dimitris. 2020. "Application of Floating Aquatic Plants in Phytoremediation of Heavy Metals Polluted Water: A Review." *Sustainability*. 12: 01-33. Accessed January 11, 2021. doi:10.3390/su12051927.

Anuradha, J., Abbasi, Tasneem and Abbasi, S. A. 2015. "An eco-friendly method of synthesizing gold nanoparticles using an otherwise worthless weed pistia (*Pistia stratiotes L.*)." *Journal of Advanced Research*. 6: 711-720. Accessed January 13, 2021. doi: http://dx.doi.org/10.1016/j.jare.2014.03.006.

Attionu, R. H. 1976. "Some effects of water lettuce (*Pistia Stratiotes, L.*) on its habitat." *Hydrobiologia*. 50: 245-254. Accessed January 12, 2021. doi: 10.1007/BF00020998.

Aurangzeb, N., Nisa, S., Bibi, Y., Javed, F., and Hussain, F. 2014. "Phytoremediation potential of aquatic herbs from steel foundry effluent." *Brazilian Journal of Chemical*. 31: 881-886. Accessed January 15, 2021. doi: 10.1590/0104-6632.20140314s00002734.

Brunel, S., Branquart, E., Fried, G., Van Valkenburg, J., Brundu, G., Starfinger, U., Buholzer, S., Uludag A., Joseffson M., and Baker R. 2010. "The EPPO prioritization process for invasive alien plants." *EPPO bulletin* 40: 407–422. Accessed January 15, 2021. Doi: 10.1111/j.1365-2338.2010.02423.x.

Burken, Joel G., Ma, Xingmao, Struckhoff, Garret C., and Gilbertson, Amanda. 2005. "Volatile organic compound fate in phytoremediation applications: natural and engineered systems." *Zeitschrift fur Naturforschung. C, Journal of Biosciences* 60: 208-215. Accessed January 11, 2021. Acess link: https://pubmed.ncbi.nlm.nih.gov/15948585/.

Cancian, Leonardo F., Camargo, Antonio F. M., and Silva, Gustavo H. G. 2009. "Growth of *Pistia stratiotes* under different temperature and photoperiod conditions." *Acta Botânica Brasilica*. 23: 552–557. Accessed January 04, 2021. doi: 10.1590/S0102-33062009000200027.

Cardoso, L. R., Martins, D., Mori, E. S. and Terra, M. A. 2005. "Genetic variability between populations of Pistia stratiotes." *Planta daninha*. 23: 181-185. Accessed January 04, 2021. doi: 10.1590/S0100-83582005000200003.

Chadwick, M. J. and Obeid, Mohammed. 1966. "A Comparative Study of the Growth of *Eichhornia Crassipes* Solms and *Pistia Stratiotes* L. at Water-Culture." *Journal of Ecology*. 54: 563-575. Accessed January 08, 2021. doi: 10.2307/2257802.

Cícero, E. A. S., Pitelli, R. A., Sena, J. A. D., and Ferraudo, A. S. 2007. "Genetic variability and sensitivity of *Pistia stratiotes* accesses to glyphosate." *Weed*. 25: 579-587. Accessed January 13, 2021. doi: 10.1590/S0100-83582007000300018.

Cilliers, Catharina. J. 1991. "Biological control of water lettuce, *Pistia stratiotes* (Araceae), in South Africa." *Agriculture, Ecosystems &*

Environment. 37: 225-229. Accessed January 04, 2021. doi: 10.1016/0167-8809(91)90151-M.

Cilliers, Catharina. J., Zeller, D., and Strydom, G. 1996. "Short- and long-term control of water lettuce (*Pistia stratiotes*) on seasonal water bodies and on a river system in the Kruger National Park, South Africa." *Hydrobiologia*, 340: 173-179. Accessed January 04, 2021. doi: 10.1007/BF00012751.

Daboor, Disse M., and Harron, Amany M. 2012. "In vitro: Potencial antimicrobiano e triagem fitoquímica de algumas plantas aquáticas egípcias." *The Egyptian Journal of Aquatic Research.* 38: 233-239. Accessed January 05, 2021. doi: 10.1016/j.ejar.2013.01.005. [In vitro: Antimicrobial potential and phytochemical screening of some Egyptian aquatic plants.]

Das, Suschimita, Goswami, Sunayana and Talukdar, Anupam D. 2014. "A Study on Cadmium Phytoremediation Potential of Water Lettuce, *Pistia stratiotes L.*." *Bulletin of Environmental Contamination and Toxicology.* 92: 169-174. Accessed January 13, 2021. doi: 10.1007/s00128-013-1152-y.

De, Anil K., Sen, Asit K., Modak, D. P., and Jana, Sasadhar. 1985. "Studies of toxic effects of Hg(II) on *Pistia stratiotes*." *Water, Air, and Soil Pollution.* 24: 351–360. Accessed January 11, 2021. doi: 10.1007 / BF00282487.

De Campos, Fernanda V., Oliveira, Juraci A., Silva, Adinan A., Ribeiro, Cleberson and Farnese, Fernanda S. 2019. "Phytoremediation of arsenite-contaminated environments: is Pistia stratiotes L. a useful tool?." *Ecological Indicators.* 104: 794-801. Accessed January 11, 2021. doi: 10.1016/j.ecolind.2019.04.048.

Dewald, L. B., and Lounibos, L. P. 1990. "Seasonal growth of *Pistia stratiotes L.* in South Florida." *Aquatic Botany.* 36: 263-275. Accessed January 12, 2021. doi: 10.1016/0304-3770(90)90040-RGet.

Di Luca, G. A., Hadad, H. R., Mufarrege, M. M., Maine, M. A., and Sánchez, G. C. 2013. "Improvement of cr phytoremediation by *Pistia stratiotes* in presence of nutrients. *International Journal of*

Phytoremediation, 16: 167–178. Accessed January 13, 2021. doi: doi:10.1080/15226514.2012.759535.

Dray Junior, F. Allen and Center, Ted D. 1989. "Seed production by *Pistia stratiotes L.* (water lettuce) in the United States." *Aquatic Botany*. 33: 155-160. Accessed January 12, 2021. doi: 10.1016/0304-3770(89)90029-6.

EPPO. 2017. *Pest risk analysis for Pistia stratiotes.* European Plant Protection Organization, Paris. 2: 01-66. Accessed January 15, 2021. Access link: https://gd.eppo.int/taxon/PIIST/documents.

Escoto, D. Fidélis, Gayer, M. Cristofary, Bianchini, M. Chimelo, Pereira, G. da Cruz, Roehrs, Rafael and Denardin, E. L. Gasparotto. 2019. "Use of *Pistia stratiotes* for phytoremediation of water resources contaminated by clomazone." *Chemosphere.* 227: 299-304. Accessed January 11, 2021. doi:10.1016/j.chemosphere.2019.04.013.

Espinoza-Quiñones, Fernando R., Módenes, Aparecido N., Costa Junior, Ismael L., Palácio, Soraya M., Szymanski, Nayara, Trigueros, Daniela E. G., Kroumov, Alexander Dimitrov and Silva, Edson A. 2009. "Kinetics of lead bioaccumulation from a hydroponic medium by aquatic macrophytes *Pistia stratiotes*." *Water, Air and Soil Pollution* 203:29-37. Accessed January 11, 2021. doi:10.1007/s11270-009-9989-8.

Farnese, Fernanda S., Oliveira, Juraci A., Gusman, Grasielle S., Leão, Gabriela A., Silveira Neidiquele M., Silva, Paulo M., Ribeiro, Cleberson and Cambraia, José. 2013. "Effects of Adding Nitroprusside on Arsenic Stressed Response of *Pistia stratiotes L.* Under Hydroponic Conditions." *International Journal of Phytoremediation.* 16: 123-137. Accessed January 11, 2021. doi: 10.1080 / 15226514.2012.759532.

Farnese, Fernanda S., Oliveira, Juraci A., Lima, F. S., Leão, Gabriela A., Gusman, Grasielle S. and Silva, L. C. 2014. "Evaluation of the potential of *Pistia stratiotes L.* (water lettuce) for bioindication and phytoremediation of aquatic environments contaminated with arsenic." *Brazilian Journal of Biology*. 74: 103-112. Accessed January 13, 2021. doi:10.1590/1519-6984.01113.

Fry, S. C. 1979. “Phenolic components of the primary cell wall and their possible role in the hormonal regulation of growth.” *Planta* 146: 343-351. Accessed January 11, 2021. doi: 10.1007/BF00387807.

Fujisawa, Takuo, Ichise, Keiko, Fukushima, Masao, Katagi, Toshiyuki and Takimoto, Yoshiyuki. 2002. “Mathematical Model of the Uptake of Non-Ionized Pesticides by Edible Root of Root Crops.” *Journal of Pesticide Science- Pesticide Science Society of Japan*. 27: 242-248. Accessed January 11, 2021. Access link:https://www.jstage.jst.go.jp/article/jpestics1975/27/3/27_3_242/_pdf/-char/ja.

Gomes, Helena I. 2012. “Phytoremediation for bioenergy: challenges and opportunities.” *Environmental Technology Reviews*. 1-8. Accessed January 11, 2021. doi: 10.1080/ 09593330.2012.696715.

Greca, M. Della, Molinaro, Antonio, Monaco, Pietro and Previtera, Lucio. 1991. “Acylglycosyl sterols from *Pistia stratiotes*.” *Phytochemistry*. 30: 2422-2424. Accessed January 12, 2021. doi: 10.1016/0031-9422(91)83671-7.

Gujarathi, Ninad P., Haney, Bryan J. and Linden, James C. 2006. “Phytoremediation Potential of Myriophyllum aquaticum and Pistia stratiotes to Modify Antibiotic Growth Promoters, Tetracycline, and Oxytetracycline, in Aqueous Wastewater Systems.” *International Journal of Phytoremediation*. 7: 99-112. Accessed January 11, 2021. doi: 10.1080/16226510590950405.

Gusain, Rita and Suthar, Surindra. 2017. “Potential of aquatic weeds (*Lemna gibba, Lemna minor, Pistia stratiotes* and *Eichhornia sp*.) In the production of biofuels.” *Process Safety and Environmental Protection*. 109: 233-241. Accessed January 11, 2021. doi: 10.1016/j.psep.2017.03.030.

Hanks, Nicole A., Caruso, Joseph A., and Zhang, Peng. 2015. “Assessing *Pistia stratiotes* for phytoremediation of silver nanoparticles and Ag(I) contaminated waters.” *Journal of Environmental Management*. 164: 41-45. Accessed January 11, 2021. doi: 10.1016/j.jenvman.2015.08.026.

Harley, Ken L. S. 1990. "Production of viable seeds by water lettuce, *Pistia stratiotes L.*, in Australia." *Aquatic Botany.* 36: 277-279. Accessed January 12, 2021. doi: 10.1016/0304-3770(90)90041-I.

Hill, Martin P. and Coetzee, Julie A. 2017. "The biological control of aquatic weeds in South Africa: current status and future challenges." *Bothalia.* 47. Accessed January 04, 2021. doi: 10.4102/abc.v47i2.2152.

Hoagland, D. R. and Arnon, D. I. 1938. "The water culture method of growing plants without soil. University of California, Berkeley." *California Agriculture Experiment Station Circular.* Accessed January 08, 2021. Access link: https://sayurankitadotcom.files.wordpress.com/2016/04/the-water-culture-method-for-growing-plants-without-soil.pdf.

Hussner, A., Heidbuechel, P., and Heiligtag, S. 2014 "Vegetative overwintering and viable seed production explain the establishment of invasive *Pistia stratiotes* in the thermally abnormal Erft River (North Rhine-Westphalia, Germany)." *Aquatic Botany.* 119: 28-32. Accessed January 13, 2021. doi: 10.1016/j.aquabot.2014.06.011.

Jarvis, S. C., Jones, L. H. P., and Clement, C. R. 1977. Uptake and transport of lead by perennial ryegrass from flowing solution culture with a controlled concentration of lead. In: *Plant and Soil.* 46: 371-379. Accessed January 08, 2021. Access link: https://link.springer.com/article/10.1007/BF00010093.

Jha, Megha, Sharma, Versha and Ganesh, Narayan. 2013. "Antioxidant and wound healing potential of *Pistia stratiotes L.*" Asian Pacific *Journal of Tropical Disease*, 2, 579–584. Accessed January 05, 2021. doi:10.1016/s2222-1808(12)60225-4.

Kaur, Manpreet, Kumar, Manoj, Sachdeva, Sarita and Puri, S. K. 2018. "Aquatic weeds as the next generation raw material for sustainable bioenergy production" *Bioresource Technology.* 251: 390-402. Accessed January 05, 2021. doi: 10.1016/j.biortech.2017.11.082.

Keates, Sarah E., Tarlyn, Nathan M., Loewus, Frank A., and Franceschi, Vincent, R. 2000. "L-Ascorbic acid and L-galactose are sources for oxalic acid and calcium oxalate in *Pistia stratiotes*." *Phytochemistry.*

53: 433-440. Accessed January 13, 2021. doi: 10.1016/s0031-9422(99)00448-3.

Koffour, George, Kyei, Samuel, Woode, Eric and Ekuadzi, Edmund. 2012. "Possible mechanism of anti-inflammatory activity and safety profile of aqueous and ethanolic leaf extracts of *Pistia stratiotes Linn* (araceae)." *Journal of Ghana Science Association*. 14: 69-81. Accessed January 05, 2021. Access link: https://www.ajol.info/index.php/jgsa/article/view/84194.

Kostman, Todd A., Franceschi, Vincent, R. Nakata, and Paul, A. 2003. "Endoplasmic reticulum sub-compartments are involved in calcium sequestration within raphide crystal idioblasts of *Pistia stratiotes L.*." *Plant Science*. 165: 205: 212. Accessed January 13, 2021. doi: 10.1016/S0168-9452(03)00160-2.

Kumar, Narendra; Jeena, Neha; Gangola, Saurabh and Singh, Hukum. 2019. Phytoremediation facilitating enzymes: an enzymatic approach for enhancing remediation process. In: Smart *Bioremediation Technologies*. 289–306. Accessed January 11, 2021. doi: 10.1016/b978-0-12-818307-6.00015-9.

Kumar, Sundeep H. K., Bose, Anindya, Raut, Arundhuti, Sahu, Sujit K. and Raju, M. B. 2010. "Evaluation of Anthelmintic Activity of *Pistia stratiotes Linn*." *Journal of basic and clinical pharmacy*. *1*: 103–105. Accessed January 05, 2021. Access link: https://www.ncbi.nlm.nih.gov/pmc/articles/PMC3979182/.

Kumar, Sundeep H. K., Raju, M. V. B., Dinda, Subas C., Sahu, S. K., and Banerjee, M. 2011. "Analgesic, anti-inflammatory and antipyretic activity of *Pistia stratiotes L.*" *Rasayan Journal of Chemistry*. 4: 506-511. Accessed January 05, 2021. Access link: https://rasayan journal.co.in/vol-4/issue-3/2.pdf.

Kumar, Vineet, Shahi, S. K., and Singh, S. 2018. Bioremediation: An Eco-sustainable Approach for Restoration of Contaminated Sites. In: *Microbial Bioprospecting for Sustainable Development,* 115–136. Accessed January 11, 2021. doi: 10.1007/ 978-981-13-0053-0_6.

Kumar Vinod, Singh, Jogendra, Saini, Ashu and Kumar, Pankaj. 2019. "Phytoremediation of copper, iron and mercury from aqueous solution by water lettuce (*Pistia stratiotes L.*)." *Environmental Sustainability*. Accessed January 11, 2021. doi:10.1007/s42398-019-00050-8.

Kumari, Alka, Lal, Brij, and Rai, U. Nath. 2016. "Assessment of native plant species for phytoremediation of heavy metals growing in the vicinity of NTPC sites, Kahalgaon, India." *International Journal of Phytoremediation*. 18: 592-597. Accessed January 11, 2021. doi: 10.1080/15226514.2015.1086301.

Kyei, Samuel, Koffuor, George A., Boampong, Johnson N. and Owusu-Afriyie, Osei. 2012. "Ocular anti-inflammatory effect of aqueous and ethanolic leaf extracts of *Pistia stratiotes Linn* (Araceae) in endotoxin-induced uveitis." *Journal of Natural Pharmaceuticals*. 3: 93-100. Accessed January 05, 2021. doi: 10.4103/2229-5119.102752.

Lahitte, Hector B., Hurrell, Julio A., Belgrano, Manuel J., Jankowski, Leandro, Haloura, Paula and Mehltreter, Klaus. 1998. Medicinal plants from the River Plate. Buenos Aires: Latin American Literature.

Lin, Yi-Li and Li, Bing-Kun. 2016. "Removal of pharmaceuticals and personal care products by *Eichhornia crassipe* and *Pistia stratiotes*." *Journal of the Taiwan Institute of Chemical Engineers*. 58: 318-323. Accessed January 11, 2021. doi: 10.1016/j.jtice.2015.06.007.

Lu, Qin, He, Zhenli L., Graetz, Donald A., Stoffella, Peter J. and Yang, Xiaoe. 2010. "Phytoremediation to remove nutrients and improve eutrophic rainwater using water lettuce (*Pistia stratiotes L.*)." *Environmental Science and Pollution Research*. 17: 84-96. Accessed January 11, 2021. doi: 10.1007/s11356-008-0094-0.

Miretzky, Patrícia, Saralegui, Andrea and Fernandez, C. Alicia. 2004. "Aquatic macrophytes potential for the simultaneous removal of heavy metals." *Chemosphere*. 57:997–1005. Accessed January 11, 2021. Doi: 10.1016/j.chemosphere.2004.07.024.

Mishima, D., Kuniki, M., Sei K., Soda, S., Ike, M. and Fujita, M. 2008. "Ethanol production from candidate energy crops: Water hyacinth (*Eichhornia crassipes*) and water lettuce (*Pistia stratiotes L.*)." *Bioresource Technology*. 99: 2495- 2500. Accessed January 11, 2021. doi: 10.1016/j.biortech.2007.04.056.

Mishra, V. Kumar and Tripathi, B. D. 2008. "Concurrent removal and accumulation of heavy metals by the three aquatic macrophytes." *Bioresource Technology*, 99: 7091–7097. Accessed January 11, 2021. doi:10.1016/j.biortech.2008.01.002.

Monaco, Pietro and Previtera, Lucio. 1991. "A steroid from *Pistia stratiotes*." Phytochemistry. 30: 2420-2422. Accessed January 12, 2021. doi: 10.1016/0031-9422(91)83670-G.

Khan, M. Azim, Marwat, K. Bahadar, Gul, Bakhtiar., Wahid, Fazli., Khan, Haroon and Hashim, Saima "*Pistia stratiotes L.* (araceae): phytochemistry, use in medicines, phytoremediation, biogas and management options." 2014. *Pakistan Journal of Botany*. 46:851-860. Accessed January 18, 2021. Access link: https://www.researchgate.net/publication/267033062_Pistia_stratiotes_L_Araceae_Phytochemistry_use_in_medicines_phytoremediation_biogas_and_management_options.

Mthethwa, Nonsikelelo P., Nasr, Mahmoud, Bux, Faizal and Kumari, Sheena. 2018. "Utilization of *Pistia stratiotes* (aquatic weed) for fermentative biohydrogen: Electron- equivalent balance, stoichiometry and cost estimate." *International Journal of Hydrogen Energy*. 43: 8243-8255. Accessed January 11, 2021. doi: 10.1016/j.ijhydene.2018.03.099.

Mthethwa, Nonsikelelo P., Nasr, Mahmoud, Kiambi, Sammy L., Bux, Faizal and Kumari, Sheena. 2019. "Biohydrogen fermentation from *Pistia stratiotes* (aquatic weed) using mixed and pure bacterial cultures." *International Journal of Hydrogen Energy*. 44: 17720-17731. Accessed January 11, 2021. doi: 10.1016/j.ijhydene.2019.05.152.

Nakata, Paul A., Kostman, Todd A. and Franceschi, Vincent R. 2003. "Calreticulin is enriched in the crystal idioblasts of *Pistia stratiotes*." *Plant Physiology and Biochemistry*. 41: 425-430. Accessed January 12, 2021. doi:10.1016/S0981-9428(03)00049-4.

Nelson, S. Glen, Smith, Barry D. and Best, Bruce R. 1981. "Kinetics of nitrate and ammonium uptake by the tropical freshwater macrophyte

Pistia stratiotes L.." Aquaculture. 24: 11-19. Accessed January 12, 2021. doi:10.1016/0044-8486(81)90039-9.

Neuenschwander, Peter., Julien, Mic H., Centre, Ted. D. and Hill, Martin P. 2009. "*Pistia stratiotes L.* (Araceae)." In: *Biological Control of Tropical Weeds Using Arthropods,* edited by Muniappan, Rangaswamy; Reddy, Gadi V. P. and Raman, Anantanarayanan. 332-352. doi: https://doi.org/10.1017/CBO9780511576348.017.

Nipaney, P. C. and Panholzer, M. B. 1987. "Influence of temperature on the biogas production of Pistia stratiotes." *Biological Wastes*. 19: 267-274. Accessed January 11, 2021. doi: 10.1016/0269-7483(87)90060-7.

Novita, V. Z., Moersidik, S. S. and Priadi, C. R. 2019. "Phytoremediation potential of *Pistia stratiotes* to reduce high concentration of copper (cu) in acid mine drainage." *IOP Conference Series: Earth and Environmental Science.* 355: 01-07. Accessed January 08, 2021. doi: 10.1088/1755-1315/355/1/012063.

Odjegba, V. J. and Fasidi, I. O. 2004. "Accumulation of trace elements by *Pistia stratiotes*: implications for phytoremediation." *Ecotoxicology.* 13:637-46. Accessed January 11, 2021. doi: 10.1007/s10646-003-4424-1.

Olguín, Eugenia J., López, Daniel A. G., Portela, Ricardo E. G., and Galván, Glória S. 2017. "Evaluation of the phytofiltration pond throughout the year using *Pistia stratiotes* in a pilot plant scale biorefinery." *Total Environmental Science*. 592: 326-33. Accessed January 11, 2021. doi: 10.1016/j.scitotenv.2017.03.067.

Pablico, P. P. and Moody, K. 1991. "Effect of fentin acetate on wet-seeded rice, *Pistia stratiotes* and *Azolla pinnata*." *Crop Protection.* 10: 45-47. Accessed January 12, 2021. doi: 10.1016/0261-2194(91)90024-L.

Pallavi, Tripathi; Sandeep, Arora; Gupta, Rajiv and Prabha, Mali R. 2011. "Diuretic activity of *Pistia stratiotes* leaf extract in rats." *International Journal Research of Pharmacy,* 2: 249-251. Accessed January 05, 2021. Access link: https://www.researchgate.net/publication/267801442_DIURETIC_ACTIVITY_OF_PISTIA_STRATIOTES_LEAF_EXTRACT_IN_RATS.

Pantawong, Rungtawan, Chuanchai, Ajcharapa, Thipbunrat, Parichat, Unpaprom, Yuwalee and Ramaraj, R. Ramaraj. 2015. "Experimental investigation of biogas production from water lettuce, *Pistia stratiotesL.*." *Emergent Life Sci Research.* 1: 41-46. Accessed January 13, 2021. Access link: https://www.researchgate.net/publication/286279142_Experimental_Investigation_of_Biogas_Production_from_Water_Lettuce_Pistia_stratiotes_L.

Pérez, M. Meza, Marín, J. César, Quintero, E. Behling, Andrade, G. Colina, Lizardo, N. Rincón and Polo, C. Arturo. 2013. "Biosorption of Pb (ii) and Cr (iii) using the aquatic plant *Pistia stratiotes." Journal of the Faculty of Engineering.* 28: 49-58. Accessed January 08, 2021. Access link: http://ve.scielo.org/scielo.php?script=sci_arttext&pid=S0798-40652013000300003.

Putra, R. Syah, Cahyana, F., and Novarita, Desi. 2015. "Removal of Lead and Copper from Contaminated Water Using EAPR System and Uptake by Water Lettuce (*Pistia Stratiotes* L.)." *Procedia Chemistry.* 14: 381-386. Accessed January 08, 2021. doi: 10.1016/j.proche.2015.03.052.

Rajkumar M., Sandhya S., Prasad M. N. V., and Freitas H. 2012. "Perspectives of plant-associated microbes in heavy metal phytoremediation." *Biotechnology Advances.* 30: 1562-1574. Accessed January 11, 2021. doi: 10.1016/j.biotechadv.2012.04.011.

Renner, Susane S., and Zhang Li, B. 2004. "Biogeography of the Pistia clade (Araceae): based on chloroplast and mitochondrial DNA sequences and Bayesian divergence time inference." *Systematic Biology* 53: 422–432. doi: 10.1080/10635150490445904.

Rezania, Shahabaldin, Din, Mohd F. M., Taib, Shazwin M., Sohaili, Johan., Chelliapan, Sohali S. C., Kamyab, Hesam and Saha, Bidyut B. 2017. "Review on fermentative biohydrogen production from water hyacinth, wheat straw and rice straw with focus on recent perspectives." *International Journal of Hydrogen Energy.* 42: 20955-20969. Accessed January 04, 2021. Doi: https://doi.org/10.1016/j.ijhydene.2017.07.007.

Ribeiro, Victor H. V., Alencar, Brenda T. B., Santos, Naiane M. C., Costa, Vitor A. M., Santos, José B., Francino, Dayana M. T., Souza, Matheus de S. and Silva, Daniel V. 2019. "Sensitivity of the macrophytes *Pistia stratiotes* and *Eichhornia crassipes* to hexazinone and dissipation of this pesticide in aquatic ecosystems." *Ecotoxicology and Environmental Safety*. 168: 177-183. Accessed January 11, 2021. Doi: 10.1016/j.ecoenv.2018.10.021.

Ricardo, Leticia M., Souza, Juliana P., Andrade, Aretha and Brandão, Maria G. L. 2017. "Plants of Brazilian Traditional Medicine: species from the books of the Polish physician Piotr Czerniewicz (Pedro Luiz Napoleão Chernoviz, 1812-1881)." *Brazilian Journal of Pharmacognosy*. 27: 388-400. Accessed January 05, 2021. doi: https://doi.org/10.1016/j.bjp.2017.01.002.

Rosa, Anne C. S., Aleixo, Matheus H. F., Harthman, Vanessa C., and Silveira, Márcio J. 2019. "Morphoanatomy of the root, stem and leaf of *Pistia stratiotes L.*" *SaBios: Health and Biology Magazine*. 14: 42-47. Accessed January 04, 2021. Access link: http://revista2.grupointegrado.br/revista/index.php/sabios2/article/view/2930

Sajna, Nina, Haler, Maja, Skornik, Sonja and Kaligaric, Mitja. 2007. "Survival and expansion of *Pistia stratiotes L.* in a thermal stream in Slovenia." *Aquatic Botany*. 87: 75-79. Accessed January 12, 2021. Doi: 10.1016/j.aquabot.2007.01.012.

Satyakala, G. and Jamil, Kaiser. 1992. "Chromium-induced biochemical changes in *Eichhornia crassipes* (Mart) solms and *Pistia stratiotes L.*." *Bulletin of Environmental Contamination and Toxicology*. 48:921-928. Accessed January 11, 2021. doi: 10.1007/BF00201155.

Service, M. W. 1962. "The possible control of Pistia stratiotes L. with diquat." *Transactions of the Royal Society of Tropical Medicine and Hygiene*. 56: 529-532. Accessed January 11, 2021. doi: 10.1016/0035-9203(62)90078-0.

Shah, Vijendra and Daverey, Achlesh. 2020. "Phytoremediation: A multidisciplinary approach to clean up heavy metal contaminated soil." *Environmental Technology & Innovation*. 18. Accessed January 11, 2021. doi: 10.1016/j.eti.2020.100774.

Sharma, B. M. and Sridhar, M. K. C. 1981. "The productivity of *Pistia stratiotes L.* in a eutrophic lake." Environmental Pollution Series A, *Ecological and Biological*. 24: 277-289. Accessed January 12, 2021. doi: 10.1016/0143-1471(81)90065-9.

Sigel, Helmuth. 1973. *Metal Ion in Biological Systems- Volume 2: Mixed-Ligand Complexes New York.*

Silva, J. V. Silva, Rosário, D. M., Veiga, A. S. S., Vasconcelos, Flávio, Percário, Sandro and Dolabela, Maria F. 2013. "A bibliographic review on the family Araceae with foccus on the genera *Pistia*, *Philodendron* and *Montrichardia*: botanical, phytochemical and biological activity aspects." *Fitos Magazine*. 8: 73-160. Accessed January 13, 2021. doi: 10.5935/1808-9569.20130006.

Silva, C. Joana da. 1981. "Observations on the reproductive biology of *Pistia stratiotes L.* (Araceae)." *Acta Amazonica*. 11: 487-504. Accessed January 12, 2021. doi: 10.1590/1809-43921981113487.

Silva Filho, Pio V. and Brandão, Mítzi. 1992. Medicinal plants of popular use collected and sold in the metropolitan region of Belo Horizonte, MG. In: *Daphne: Herbarium Magazine Agricultural Research Corporation of Minas Gerais*. Belo Horizonte: EPAMIG.

Sinang, Som C., Daud, Norhayati, Kamaruddin, Nurhaida and Poh, Keong B. 2019. "Potential for inhibiting the growth of freshwater algae by extracts of herbaceous plants." *Acta Ecologica Sinica*. 39: 229-233. Accessed January 08, 2021. doi: https://doi.org/10.1016/j.chnaes.2018.12.005.

Sinha, Sarita; Saxena, Rohit and Singh, Shraddha. 2005. "Chromium induced lipid peroxidation in the plants of *Pistia stratiotes L.*: role of antioxidants and antioxidant enzymes." *Chemosphere*. 58: 595-604. Accessed January 13, 2021. doi: https://doi.org/10.1016/j.chemosphere.2004.08.071.

Sridhar, M. K. C. and Sharma, B. M. 1985. "Some observations on the oxygen changes in a lake covered with *Pistia stratiotes L.*." *Water Research*. 19: 935-939. Accessed January 12, 2021. doi:10.1016/0043-1354(85)90155-1.

Sridhar, M. K. C. 1986. "Trace element composition of *Pistia stratiotes L.* in a polluted lake in Nigeria." *Hydrobiologia*. 131: 273-276. Accessed January 11, 2021. doi: 10.1007/BF00008863.

Sudiarto, S. I. Amalia, Renggamen, Anriansyah and Choi, H. Lim. 2019. "Floating aquatic plants for total nitrogen and phosphorus removal from treated swine wastewater and their biomass characteristics." *Journal of Environmental Management*. 231: 763-769. Accessed January 11, 2021. doi: 10.1016/j.jenvman.2018.10.070.

Tavares, K. Sendra. 2003. "The aquatic macrophyte community in reservoirs in the middle and lower Tietê river (SP) and in lakes in the middle Rio Doce basin (MG)." Master degree, University of São Carlos. Accessed January 11, 2021. Access link: https://repositorio.ufscar.br/bitstream/handle/ufscar/2060/DissKST.pdf?sequence=1&isAllowed=y.

Tabinda, A. Bari. Irfan, Rabia, Yasar, Abdullah, Iqbal, Anum and Mahmood, E. Adeel. 2020. "Phytoremediation potential of *Pistia stratiotes* and *Eichhornia crassipes* to remove chromium and copper." *Environmental Technology*. 41: 1514-1519. Accessed January 11, 2021. doi: 10.1080/09593330.2018.1540662.

Tarlyn, Nathan M., Kostman, Todd A., Nakata, Paul A., Keates, Sarah E., and Franceschi, Vincent, R. 1998. "Axenic culture of *Pistia stratiotes* for use in plant biochemical studies." *Aquatic Botany*. 60: 161-168. Accessed January 13, 2021. doi: ttps://doi.org/10.1016/S0304-3770(97)00078-8.

Tewari, Anamika, Singh, Ragini, Singh, Kumar N. and Rai, U. N. 2008. "Amelioration of municipal sludge by *Pistia stratiotes L.*: Role of antioxidant enzymes in detoxification of metals." *Bioresource Technology*. 99: 8715-8721. Accessed January 11, 2021. doi: 10.1016/j.biortech.2008.04.018.

Toursarkissian, Martin. 1980. *Medicinal plants of Argentina: their botanical and common names, uses and geographical distribution*. Buenos Aires: Editorial Hemisferio Sur.

Tripathi, S., Singh, V. K., Srivastava, P., Singh, R., Devi, R. S., Kumar, A. and Bhadouria, R. 2020. Phytoremediation of organic pollutants.

Abatement of Environmental Pollutants. In Abatement of Environmental Pollutants; *Trends and Strategies*. 81–105. doi:10.1016/b978-0-12-818095-2.00004-7.

Tyagi, T. and Agarwal, M. 2017. "GC-MS analysis of invasive aquatic weed, *Pistia stratiotes* L. and *Eichhornia crassipes* (mart.) solms." *International Journal of Current Pharmaceutical Research*. 9: 111-117. Accessed January 13, 2021. doi:10.22159/ijcpr.2017.v9i3.19970.

Upadhyay, R. K. and Panda, S. K. 2005. "Salt tolerance of two aquatic macrophytes, *Pistia stratiotes* and *Salvinia molesta*." *Biologia Plantarum*. 49: 157-159. Accessed January 13, 2021. doi:10.1007/s10535-005-7159-4.

Upadhyay, R. K. and Panda, S. K. 2009. "Copper-induced growth inhibition, oxidative stress and ultrastructural alterations in freshly grown water lettuce (*Pistia stratiotes L.*)." *Comptes Rendus Biologies*. 332: 623-632. Accessed January 13, 2021. doi: 10.1016/j.crvi.2009.03.001

Wang C. J. and Liu, Z. Q. 2007. "Foliar uptake of pesticides—present status and future challenge." *Pesticide Biochemistry and Physiology*. 87: 1–8. Accessed January 13, 2021. doi: 10.1016/j.pestbp.2006.04.004.

Wang, Jinqing, Fu, Guihua, Li, Weiyue, Shi, Ying, Pang, Jicai, Wang, Qiang, Lü, Weiguang, Liu, Change and Liu, Jiansheng. 2018. "The effects of two free-floating plants (*Eichhornia crassipes* and *Pistia stratiotes*) on the burrow morphology and water quality characteristics of pond loach (*Misgurnus anguillicaudatus*) habitat." *Aquaculture and Fisheries*. 3: 22-29. Accessed January 13, 2021. doi:10.1016/j.aaf.2017.12.001.

Wei, Zihan, Quyet, Van L., Peng, Wanxi, Yang, Yafeng, Yang, Han, Gu, Haiping, Lam, Su S. and Sonne, Christian. 2021. "A review on phytoremediation of contaminants in air, water and soil." *Journal of Hazardous Materials*. 403. Accessed January 11, 2021. doi:/10.1016/j.jhazmat.2020.123658.

Winton, Mary D. and Clayton, John S. 1996. "The impact of invasive submerged weed species on seed banks in lake sediments." *Aquatic*

Botany. 53: 31-45. Accessed January 04, 2021. doi:10.1016/0304-3770(95)01010-6.

Wiszniewska, Alina, Hanus-Fajerska, Ewa, Muszynska, Ewa, and Ciarkowska, Krystyna. 2016. "Natural organic amendments for improved phytoremediation of polluted soils: a review of recent progress." *Pedosphere.* 26: 1–12. Accessed January 13, 2021. doi: 10.1016/S1002-0160(15)60017-0.

Yu-Chen, Kuan, Zheng, Yi and Cheng, Yu Cheng. 2015. "Integrated alkali pretreatment and preservation of wet lettuce (*Pistia stratiotes*) by lactic acid bacteria for fermentable sugar production." *Biomass and Bioenergy.* 81: 249-255. Accessed January 13, 2021. doi: 10.1016/j.biombioe.2015.07.007.

Zimmels,Y., Kirzhner F. and Malkovskaja A. 2006. "Application of *Eichhornia crassipes* and *Pistia stratiotes* for treatment of urban sewage in Israel." *Journal of Environmental Management* 81: 420–428. Accessed January 11, 2021. doi: 10.1016/j.jenvman.2005.11.014.

Zennie, Thomas M. and Mcclure, Jerry W. 1977. "The flavonoid chemistry of *Pistia stratiotes L.* and the origin of the lemnaceae." *Aquatic Botany.* 3: 49-54. Accessed January 12, 2021. doi: 10.1016/0304-3770(77)90003-1.

In: The Future of Phytoremediation
Editor: James C. Flores
ISBN: 978-1-53619-625-2

Chapter 4

PHYTOREMEDIATION OF DAIRY WASTEWATER BY CONSTRUCTED WETLAND USING MACROPHYTES: TECHNICAL AND ENVIRONMENTAL ASPECTS

Rita de Cássia Souza de Queiroz[1], José Adolfo de Almeida Neto[2] and Luciano Brito Rodrigues[3,*]

[1]Department of Education, Federal Institute of Bahia, Jequié, Bahia
[2]Department of Agricultural and Environmental Sciences, State University of Santa Cruz, Ilhéus, Bahia
[3]Department of Rural and Animal Technology, State University of Southwest Bahia, Itapetinga, Bahia

ABSTRACT

The industries seek innovative processes to treat and valorize their by-products, residues, and wastewater, considering the economic needs,

* Corresponding Author's E-mail: rodrigueslb@uesb.edu.br.

the growing socio-environmental demands, and the stricter environmental legislation. The dairy industry products are responsible for relevant environmental aspects, mainly concerning energy consumption and wastewater' generation related to its life cycle. In this sense, mitigating measures should also be investigated, considering the significant number of industries and the economic and social relevance of this activity in Brazil and worldwide. The treatment and recovery of by-products, waste, and effluents within a perspective of process integration and industrial ecology are current trends. The recovery allows the cycling of nutrients and by-products, residues and wastewater in new production stages. However, the high number of small and medium industries, associated with the reduced volume of effluent generated by each company and their geographic distribution, are challenges for its application in the dairy industry, mainly related to the transport and perishability of these products. In this perspective, studies highlight the use of constructed wetlands as an alternative in the treatment and recovery of dairy wastewater. Compared to other treatment systems, the constructed wetlands have advantages such as ease of operation, a significant reduction in the environmental impact, and implementation and maintenance costs. The main component of the constructed wetlands are aquatic macrophytes, and their biomass can be used after treatment. The choice of aquatic plants must be judicious for developing the plants and maintaining the wastewater treatment. The plant's local occurrence will also indicate its adaptation to the local climate, being one of the essential factors for the development and maintenance of macrophytes during treatment. Different aquatic macrophytes are successfully used in different regions to treat domestic and industrial wastewater and provide nutrient cyclings. This chapter will address the technique of treating industrial wastewater through constructedf wetlands using macrophytes, highlighting the historical aspects, the main characteristics, and influencing factors, management, and challenges for using such technique, considering its environmental performance of nutrient cycling from the dairy industry. The results indicate that the application of constructed wetlands is promising, increasing the potential for use, both on a large and small scale, and maybe an appropriate method for application in small and medium dairy industries.

Keywords: industrial ecology, phytoremediation, constructed wetlands, Life Cycle Assessment, Clean Technologies

INTRODUCTION

Considering economic needs, growing socio-environmental demands and stricter environmental legislation, industries have demanded innovative processes in treating and valorizing their by-products, waste, and effluents.

Regarding the life cycle of dairy products, the highest levels of negative environmental impact in the chain are mostly from farms, mainly due to greenhouse gas emissions. Such impacts are the most studied for mitigation in the milk chain. Dairy industries can also responsible for relevant environmental problems, especially concerning energy consumption and water resource pollution.

Thus, mitigating environmental impacts' actions must be evaluated due to the number of dairy industries and their social-economic relevance in Brazil and worldwide.

The demand for dairy products shows significant increase projections and industrialization rates to supply the population growth. Such factors require greater consumption of raw materials and inputs, such as water and energy, increasing waste generation.

This situation contributes to the emergence of environmental problems, such as the improper disposal of wastewater. This disposal causes associated nutritional waste and eutrophication with the reduction of dissolved oxygen in water. Thus, three environmental indicators stand out for dairy products: water, energy consumption, and effluents' generation and disposal. The second indicator is more relevant in smaller industries that spending the same amount of energy to process smaller volumes of milk when compared to the larger ones.

The treatment and valorization of by-products, residues, and effluents within a process integration perspective and using industrial ecology principles are a current trend.

The dairy industries appear as a sector with potentials of applying these concepts due to the properties of their wastewater. However, applying these concepts in the dairy industries faces some challenges, making the integration of processes unfeasible. The small volume of wastewater generated individually and the high number of medium and small dairy far away geographically requires the transport of these effluents, high perishability.

Lin et al. (2014) pointed out that dairy wastewater management has been limited mainly to the practice of disposal in the environment, even though it is a valuable source of organic matter and energy. Thus, the environmental impact around the dairy can be significant, resulting in severe damage to watersheds, interfering in water quality for consumption and leisure.

The forms of recovery for dairy wastewater mainly used are:

- membrane separation processes, by removing specific components of the wastewater, which can lead to water reuse (Wallace and Safferman 2014);
- fermentations by generating new products such as ethanol, biogas, and lactic acid (Carvalho, Prazeres, and Rivas 2013) and,
- the constructed wetlands using macrophytes, with the possible usage of their biomass (Gorra et al. 2014).

Generally, treatment and recovery technologies have high implementation and maintenance costs that are not compatible with medium and small industries' revenue, even the most conventional treating technologies. On the other hand, studies have highlighted the implementation of constructed wetlands as an attractive system for wastewater treatment, both in the environmental and financial aspects. Lee et al. (2010) emphasized the reduced cost for implementing and maintaining constructed wetlands compared to other forms of wastewater treatment. Matamoros, García, and Bayona (2008) steted that the use of wetlands to reduce pollutants and contaminants in wastewater reached tertiary treatment levels, comprising high costs. In addition to reducing various contaminants, such as organic and inorganic loads, toxic metals, pesticides, and pathogenic microorganisms, the constructed wetlands do not need extensive areas and promote an overview of process integration through the use of biomass and nutrient cycling (Carvalho, Prazeres, and Rivas 2013; Matamoros, García, and Bayona 2008).

The main component of the constructed wetlands are aquatic macrophytes, which are living organisms which require careful choice for the success and maintenance of the structure. Bhatia and Goyal (2014) and Martelo and Borrero (2012) pointed out the following criteria: type of plant used (development and maintenance ratio with efficiency in removing pollution), seasonal growth (reduction of efficiency during winter), and whether the species occurs close to the treatment site. According to its incidence in the region, the plant's choice, besides showing availability, indicates adaptation to the local climate. Such factors are critical for developing and maintaining macrophytes during their use in the wetland.

Different aquatic macrophytes are being used to treat domestic and industrial wastewater, providing nutrient cyclings, such as nitrogen and phosphorus. (Carvalho, Prazeres, and Rivas (2013) addressed the implementation of constructed wetlands to wastewater treatment from small and medium-sized dairy industries, including rural ones.

Mendonça et al. (2012) showed a lack of studies on dairy effluents' treatment with constructed wetlands in a tropical climate. Lu et al. (2016) investigated different impacts of constructed wetlands in rural areas. The authors suggested further experimental studies on the different conditions that should be considered in wetland treatments, such as operation, flow, and types of materials used to support the plants.

In this sense, this chapter seeks to demonstrate how the system of constructed wetlands with local aquatic macrophytes is a relevant technology in treating wastewater, both under environmental and technical issues.

1. Constructed Wetlands

Constructed wetlands (CWs) are "wetlands" that use aquatic macrophytes to reduce organic and inorganic matter, as well as allowing the removal of microorganisms (including pathogens), toxic metals and promoting nutrient cycling through the use of vegetable biomass.

CWs comprise a system considered as emergent, which has been identified as a solution for the valorization of rainwater, domestic sewage, ponds, and rivers already eutrophic, industrial, and agricultural wastewater (Bhatia and Goyal 2014; Golda, Poyyamoli, and Nandhivarman 2014; Kadlec and Wallace 2008). The CWs can be used worldwide in the most diverse sectors, including small and medium-sized dairy industries (Queiroz et al. 2019).

Shaw and Fredine (1956) were probably the first authors to use the nomenclature of wetlands. The authors define them as shallow, temporary, or intermittent plains, with different names such as swamps, wetlands, marshes, and lands where the river overflows with emergent vegetation. The definition does not include rivers, lakes, and temporary waters that they do not develop moist-soil vegetation.

When the nomenclature is used, CWs have varieties, such as constructed flooded systems (Mendonça et al. 2015), flooded areas (Diniz et al. 2005), root zones (Schirmer and Oliveira 2010), and filters planted with macrophytes (Pelissari et al. 2013).

CWs are considered biological systems that mimic the natural environment. Thus, to achieve their objectives is essential the interaction between their elements, i.e., the type of water or effluent used, the aquatic macrophytes, the support medium (soil, gravel, gravel, or artificial holder), microorganisms (from the roots of vegetation) and also from the soil (Lin et al. 2005). It is also necessary to consider climatic fluctuations, solar incidence, hydraulic retention time, temperature, and pH of the medium to the detriment of the system's performance. It is emphasized that the water or effluent's quality is one of a CW system's primary purposes.

1.1. Advantages and Disadvantages

Research shows that CWs have an advantage compared to other treatment systems (Hubbard 2010; Wu et al. 2014; Zhang et al. 2012;

Vymazal 2009; Carvalho, Prazeres, and Rivas 2013; Ansari and Gill 2014; Bhatia and Goyal 2014; Gorra et al. 2014b; Sultana et al. 2015), such as:

- Low costs for building, maintenance, and service;
- Not require great depths for use in lagoons or canals;
- Easy operation;
- Environmentally viable solution;
- High potential for treating different sewages;
- Potential for nutrient cycling;
- Promotes the integration of process and use of biomass;
- Reduction in the values of organic, inorganic matter, toxic metals, pesticides, and pathogenic microorganisms;
- Possibility of associating macrophytes for greater treatment efficiency;
- Growth of macrophytes throughout the year in tropical to subtropical climates.

The growth rates of aquatic plants are relevant and fast. Their biochemical content depends, mainly, where the biomass develops (Saqib et al. 2013). The consortium or association among different species can help achieve more efficient results in removing organic and inorganic material (Ansari and Gill 2014). On the other hand, the competition between species can reduce the system's efficiency, requiring some management care.

The CWs disadvantages include low efficiency with coarse particles and mosquitoes' possible proliferation, mainly for horizontal surface flow systems, also called free water systems. In such systems, the wastewater is exposed to the environment, requiring large construction areas to treat large volumes (Headley and Tanner 2012; Martelo and Borrero 2012). However, Queiroz et al. (2020) observed that previous stages of grating and filtration before CWs resulted in decreased solids content and increased efficiency during phytoremediation. Another challenge for using CW systems is the appropriate final biomass disposal (Bhatia and Goyal

2014). The biomass should be used as a raw material in new processes, closing nutritious and energetic cycles according to industrial ecology and circular economy principles.

The cost-benefit of CWs has been considering its evolution with the adaptation to sustainable development activities. The sustainability indicators most used in the cost-benefit analysis for CWs comprise land acquisition and use, energy consumption, investment and operating costs, and ecological benefits (Wu et al. 2015). Land use can be a limiting indicator for treatment with CW.

1.2. Classification of Constructed Wetlands

The classification of CWs can range from the type of flow in the system (i.e., superficial, horizontal sub-surface, and vertical sub-surface) and the variety of aquatic macrophytes (Austin 2006). The main classification concerning plants divides them into a) emergent or emerged aquatic macrophytes: plants rooted in the soil and with leaves out of the water; b) submerged aquatic macrophytes: plants that may or may not be rooted in the soil; and c) floating aquatic macrophytes: are those that float on the surface of the water (Esteves 1998). Figure 1 illustrates the types of wetlands.

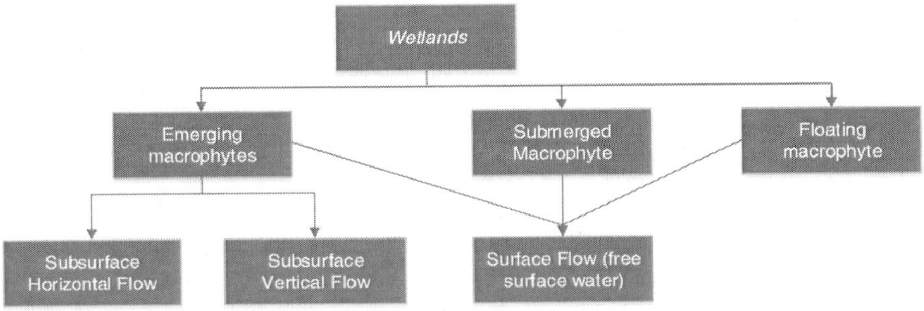

Source: own authorship.

Figure 1. Types of treatment with wetlands according to the aquatic macrophyte used.

Emerging aquatic macrophytes attached to supports are being used for phytoremediation. Such arrangement is called floating constructed wetlands or floating treatment wetland technologies. Besides its low cost and easy operation and maintenance, constructed wetlands with free surface water and emerging floating macrophytes can treat larger volumes when compared to other CW systems. Such systems do not need supporting material (gravel and soil, for example), and the roots have more contact with the effluent to be treated (Headley and Tanner 2012). Cronk (1996) and Ladislas et al. (2015) also indicated positive effects on the filtration in the roots of plants emerging in floating wetlands with free surface water.

Bhatia and Goyal (2014) indicated that the large-scale use of wetland systems requires research at a fundamental and applied level, seeking to overcome the limits and improve levels of removals of organic and inorganic compounds to seasonality. According to the season, the limiting factors are related to the systems' removal efficiency, biomass disposal, CW loading limits, and plant growth variation. As the studies advance, the authors also point out that multiple purposes will be achieved, such as wastewater treatment, biomass reuse, and environmentally friendly approaches. An earlier contribution by Hobbie (1992) addresses that plants can create nutrient cycling patterns in natural environments.

1.3. Aquatic Macrofites in the Treatment, Recovery, and Cycling of Nutrients in CWs

The use of different aquatic macrophytes has gained prominence in treating effluents, mainly due to the ease of cultivation and harvesting, compared to microalgae. They have demonstrated success in treating different effluents, e.g., the domestic and industrial ones, providing nutrient cyclings, such as nitrogen and phosphorus.

Features as ease of adapting and growing in different cultivation and media conditions, reducing organic and inorganic load, and the CO2 capture are other significant aquatic macrophytes' attributes (Aresta, Dibenedetto, and Barberio 2005).

Macrophytes are useful in the wastewater treatment. After its use, the plants can yields biomass for energy and fertilizer production (Pompêo 2008; Saqib et al. 2013).

The choice of the proper aquatic macrophytes is crucial once the species is directly related to the efficiency in removing organic and inorganic loads. Such choice must also consider the seasonal growth due to the reduction in efficiency cycling during winter. The species' availability close to CWs place and the relating of effluent flow with the hydraulic retention time are also essential in the macrophyte's choice (Bhatia and Goyal 2014; Martelo and Borrero 2012).

Weaver and Clements in 1938 suggested a definition for aquatic macrophytes as herbaceous plants that grow in water, in covered or water-saturated soils. According to Esteves (1998), the authors defined the term very broadly, encompassing vegetables that inhabit swamps to aquatic environments, ranging from macroalgae to angiosperms, such as those of *Chara* and *Typha* genera.

Current literature describes several species of aquatic macrophytes used for phytoremediation purposes in CWs. Bhatia and Goyal (2014) compared studies with phytoremediation processes in different wastewater and concluded that Phragmites is the most used genus in wetland systems, representing 41%. The other macrophyts are Typha 23%, Spartina 15%, Scirpus 8%, and Juncus 5%. The Acorus, Cyperus, and Lemna are examples of the less used (Bhatia and Goyal 2014).

1.4. Mechanisms for Organic and Inorganic Material Reduction in WCS

The main mechanisms in the treatment using CWs involve physical, chemical, and biological processes. Mendonça et al. (2015) highlighted some examples, such as filtration, microbial degradation, absorption, and adsorption. Microorganisms are essential in reducing pollutants and contaminants. The CWs provide an optimal environment for their development, degradation, and assimilation of organic and inorganic compounds. They act in the primary treatment in CWs, contributing to the absorption, transformation, and recycling processes of components present in different wastewater (Kadlec and Wallace 2008).

The characteristic of removing pollutants and contaminants using aquatic plants and microorganisms contained in their roots is called phytoremediation. The CW is the technology designed for this purpose (Dipu, Kumar, and Thanga 2011). The removal of organic matter occurs through microorganisms, especially aerobic ones, which use in their metabolism oxygen dissolved in water or effluent and oxygen released by the roots of macrophytes to oxidize the organic matter in the effluent (Von Sperling 1996). During the treatment, the microorganisms come from the plants' rhizomes and the tributary itself, forming a microbial biofilm able to carry out this degradation. Von Sperling (1996) points out that as long as organic matter and oxygen are available, there will be the development of microorganisms, converting organic matter into inert and straightforward compounds according to the reaction:

organic matter + O2 + microorganisms → CO2 + H2O + microorganisms + -energy.

Headley and Tanner (2012) state that particulate or suspended organic matter is removed mainly by sedimenting and filtering performed by the aquatic macrophytes' roots.

The removal of nitrogen in CWs occurs mainly through nitrification and denitrification. The before-mentioned processes are related to organic nitrogen, in which absorption by plants and storage in sediments are the main factors in regulating nitrogen removal (Ma et al. 2016). The nitrification stage begins with ammonium being oxidized to nitrite ($NH4^+ + 1.5O_2 \rightarrow NO_2^- + 2H^+ + H_2O$) and nitrite oxidized to nitrate ($NO_2^- + 0.5O_2 \rightarrow NO_3^-$). Such processes are conducted by bacteria of the genera Nitrosomonas and Nitrobacter in the presence of oxygen, and the direct reaction is as follows, $NH_4^+ + 2O_2 \rightarrow NO_3^- + 2H^+ + H2O$ (Vymazal 2007).

The denitrification process comprises the nitrate reduction to nitrogen gas. The process requires anoxic conditions and denitrifying bacteria to occurs (Ma et al. 2016; Wang et al. 2016). Wang et al. (2016) expressed the global denitrification equation as $NO_3^- + 10e^- + 12H^+ \rightarrow N_2 + 6H_2O$.

Sirivedhin and Gray (2006) researched the factors that influence the denitrification process. They concluded that among the factors of hydraulic conditions (i.e., depth and aeration), microbial community, nitrate concentration, and quantity and quality of organic carbon from the affluent, the last factor was indicated one for supplying electron donors for the denitrification process.

In CWs, phosphorus occurs as phosphate in organic and inorganic forms. Free orthophosphate is the form assimilated directly by the algae and aquatic macrophytes. Following it is converted into structural organic phosphorus and stored as biomass in vegetation (Vymazal 2007).

The absorption by the root system of plants and microbial degradation processes of organic phosphorus are the primary mechanisms for removing this element (Ma et al. 2016). Vymazal (2007) highlights that for the quantification of phosphorus contents in CWs it is essential to evaluate phosphorus absorbed by plants and accumulation in sediments, uptake in microbial biofilm adsorption, desorption, and leaching. Since the transformation of phosphorus in these environments occurs in all these mechanisms. On the other hand, phosphorus sorption and its biomass storage are saturable (Dunne, Reddy, and Carton 2005). Thus, the plant's saturation period for collecting aquatic macrophytes and the phosphorus removal system's sustainability must be understood, preventing the returning to the CWs through leaf fall and plant decomposition.

2. Constructed Wetlands and the Milk Sector

2.1. Dairy Wastewater: Generation and Treatment

Dairy wastewater, also known as dairy wastewater (DW), can be composed of cleaning and washing water, detergents, cooling water, wasted milk and derivates, equipment lubrication material, and sanitary waste (CETESB 2008). Whey can be part of the composition when added to other sewages.

The amount of wastewater generated in dairy industries is directly related to water consumption, the cleaning process, and the whey produced. The closed system hygiene method, named Clean in Place (CIP), reduces water volume during the washing of pipes and equipment in dairies industries. Generally, the quantity ranges from 1 to 5 liters of effluent produced per liter of milk processed (Matos et al. 2010).

In cheese production, around 4 liters of wastewater are generated per liter of processed milk (Carvalho, Prazeres, and Rivas 2013). A survey that evaluated water consumption and wastewater generation in a dairy industry in Brazil concluded that the wastewater generation coefficient is 3.5 liters for each liter of processed milk (Saraiva et al. 2009).

The mentioned data indicate the expressive wastewater generation compared to the total volume of milk received daily in the dairy industry. However, Lin et al. (2014) addressed that this effluent's management has been limited to disposal practices worldwide. Thus, it is necessary to apply techniques for the treatment and or recovery of the dairy wastewater after its generation. The forms of treatment for dairy wastewater include biological processes (i.e., aerobic and anaerobic degradation), physical-chemical (i.e., precipitation and oxidation treatments), mechanical and also the constructed wetlands (Carvalho, Prazeres, and Rivas 2013; Slavov 2017). The authors also highlight the importance of previous recovering of the whey components considering their high nutritional value.

In general, the wastewater treatments have some disadvantages that can be the key to their application. These disadvantages are the process instability, non-removal of COD and BOD rates at acceptable levels for discharge into the natural environment, and the need to integrate with other processes, requiring long retention and large area (Rivas, Prazeres, and Carvalho 2011). Mendonça et al. (2012) assure that conventional treatment plants cannot promote conditions for effective removal of phosphorus or the microbiological processes of nitrification and denitrification to remove nitrogen. Since anaerobic digestion does not remove nitrogen and phosphorus, other technologies must be employed to reduce their concentrations to legal discharge limits (Milani, Nutter, and Thoma 2011).

Most of the current treatment systems do not allow the direct discharge of effluents into water. Besides, medium and small-sized dairy industries are dispersed and with a volume of wastewater that limits more expensive processes. This condition requires implementing more than one technology and or integrated processes for treatment to discharge into water (Carvalho, Prazeres, and Rivas 2013). However, the authors do not indicate the system's real values, so there is no clarity on which makes the processes costly.

2.2. Environmental Problems of the Milk Sector

It is necessary to increase the consumption of raw materials and inputs (for example, water, energy, ingredients) to guarantee the supply and growing demand for industrialized products, increasing the generation of waste and negative environmental impacts in all product's life cycle.

Capper (2013) warns that the increase in world production of animal origin products, to supply the demand in the next 40 years, cannot lose sight of the issue of sustainability.

The food chain is considered responsible for 21% of the total anthropogenic emission of greenhouse gases. In this percentage, 2.7% is due to the milk chain's contribution worldwide (Milani, Nutter, and Thoma 2011; Toma et al. 2013). Such values are examples of just one category of environmental impact. So, it demonstrates the need for studies that contribute to the identification, evaluation, and mitigation of the impacts caused by the food sector in order to mitigate its environmental outputs. The concept of an organization's environmental performance can be attributed to the measurable result of management on environmental aspects, which represent elements of an establishment's activities or products that can interact with the environment and cause environmental impact (ABNT 2004).

Considering the milk chain, the farm is responsible for most of significant adverse environmental effects. The most outstanding impact categories on dairy farm studies are global warming and land use (Berlin, Sonesson, and Tillman 2007; Fantin et al. 2012; Finnegan et al. 2015; González-García et al. 2013; Yan, Humphreys, and Holden 2011; Zehetmeier et al. 2013). Several studies are aiming to improve the environmental performance of dairy farms. However, to achieve success in dairy products' life cycle, improvements must cover all chain stages. So, the environmental performance of dairy industries should also be studied. These industries are responsible for adverse environmental impacts, mainly when their residues, effluents, and emissions are not adequately managed. The prime negative consequences are pollution and eutrophication of river waters around the industry. There is also nutritional waste and the need to assess how dairy wastewater is treated, which causes high environmental impacts.

In dairy industries, the environmental impacts vary according to what is produced, the type of technology, and the level of automation used. In dairy industries that process several products, Djekic et al. (2014) pointed out the demand for energy and input as causing the main environmental problems. Xu and Flapper (2009) pointed out global warming and eutrophication. González-García et al. (2013) indicated energy and water consumption, with consequent generation of high amounts of effluents. Høgaas Eide (2002) concluded that small dairy industries are responsible for a higher environmental impact than medium and large dairy, mainly for the energy consumption indicator. These industries require the same energy for processing small volumes.

The main negative environmental impact of dairy industries in Brazil occurs due to the discharge of untreated wastewater into the environment, being an evident practice, mainly in small dairy industries (Barbosa, Florentino, Florencio 2010; Saraiva et al. 2009).

Pollution caused by dairy wastewater' disposal is related to its high organic load and high nutritional composition (Table 1).

Table 1. Physical-chemical characterization of the dairy wastewater according to the literature

Authors	BOD	COD	OD	N	P	TS	SD	SS
Brião and Tavares (2007)	-	2,491	-	69.4	37.5	-	-	-
(Dipu, Kumar, and Thanga 2011)	1,300	2,920	-	-	-	1,986	1,378	6.44
(Geraldino 2014)	4,670	11,817	4.08	-	-	-	-	-
(Gorra et al. 2014a)	725	-	-	198	10	-	-	-
(Matos et al. 2010)	2,790	5,143	-	89.5	22.5	2,407	1,705	702

Parameters expressed in mg L^{-1}.

BOD: biochemical oxygen demand; COD: chemical oxygen demand; DO: dissolved oxygen; N: nitrogen; P: phosphorus; TS: total solids; DS: dissolved solids; SS: suspended solids.

Table 2. Difference in the dairy effluent's characterization according to the product processed in the dairy

Processed products	BOD	COD	SS	ST	SSed	NT	TP
Pasteurized milk and butter	487-1,319	2,491	-	69.4	37.5	-	-
Pasteurized milk and yogurt		2,920	-	-	-	1,986	6.44
Sterilized milk and yogurt	4,670	11,817	4.08	-	-	-	-
Condensed milk	725	-	-	198	10	-	-
Powdered milk	2,790	5,143	-	89.5	22.5	2,407	702

Adapted from CETESB (1990)

Parameters expressed in mg L^{-1}.

BOD: biochemical oxygen demand; COD: chemical oxygen demand; SS: suspended solids; TS: total solids; SSed: sedimentable solids; TN: total nitrogen; TP: total phosphorus.

These compounds are carried through the factory's cleaning and sanitization processes, presenting a significant physical-chemical characteristics variation. Such characteristics depend on a series of aspects, emphasizing products processed in the dairy or production lines (Table 2), the technology used, hygiene procedure (Rivas et al. 2010). Moreover, the company's size is also important since most of large dairy industries reuse whey in new products and recover the effluent's chemical components.

Høgaas Eide (2002) calculated that 80% of eutrophication impacts and 30% of energy use come from cleaning and sanitizing of dairy industries.

The dairy effluent's organic load is mainly related to protein compounds, carbohydrates, and lipid fraction (Silva et al. 2013). Its removal can be observed by the decrease in Chemical Oxygen Demand (COD) and Biochemical Oxygen Demand (DBO).

Nutrients such as nitrogen (N), phosphorus (P), calcium (Ca), potassium (K), and sodium (Na) present in dairy wastewater have similar or often higher levels than domestic sewage (Dipu, Kumar, and Thanga 2011; Mendonça et al. 2012). As much as they represent critical elements to life, such nutrients contribute to eutrophication and water quality changes if discarded in the water (Ibekwe, Grieve, and Lyon 2003).

The eutrophication phenomenon occurs when water bodies receive a high concentration of nutrients, mainly N and P. Such release triggers an excessive growth of algae and cyanobacteria, followed by oxygen depletion and consequent mortality of aquatic fauna and flora. Ansari and Gill (2014) highlight that eutrophication was noticed as a problem in the middle of the 20th century, and then the studies intensified to avoid its worsening. Besides modifying the ecological structure of water, eutrophication changes its functionality (leisure, use in agriculture and supply), reducing its quality and increasing treatment costs (Dodds et al. 2009; Quilliam et al. 2015). The removal of N depends on oxidative processes of ammonia in nitrite or nitrate, which is carried out by microbial processes (Austin 2008). The soluble P is assimilated by the plant and converted into structural organic P, being stored in the biomass of aquatic macrophytes and algae (Mendonça et al. 2012).

Baker (2011) considered that P's world stocks' availability and the high extraction rates might promote a scarcity of this element. Vaccari et al. (2014) point out that the scarcity may not be related to this element's quantity in nature. It is also necessary to consider the difficulty of accessing reserves and technologies for the extraction or the competition of land use, since this will be more valuable in other uses, preventing the deposits' economic exploitation. Jarvie, Neal, and Withers (2006) state that

P's significant releases in the rivers come from wastewater, many types of industrial wastewater, and agriculture.

Rockström et al. (2009) evaluated the Earth's resilience capacity against human actions, leading to irrevocable environmental problems. O, N and P were among the nine parameters studied and the seven with the quantitative limits established. In 2015, Rockström published the planetary limits update and concluded that 4 out of 9 categories analyzed exceeded the established limits. Such categories were classified as a "high-risk zone" for the integrity of the biosphere and biogeochemical cycles (P and N) and as a "danger zone" for climate change and land use (Rockström 2015).

Given the above, the productive cycles of these elements must be closed, avoiding their loss with the improper disposal of effluents. Such initiative would reduce P's extraction and, consequently, reduce competition for land use and prevent river eutrophication and nutritional waste.

2.3. Principles of Industrial Ecology Applied to the Milk Sector

2.3.1. Valorization and Cycling of Nutrients from Dairy Wastewater and Company Size

Several ways of "waste recovery"[1] are being used in industries, mainly to recover compounds, such as Membrane Separation Process (PSM), fermentations, systems with micro or macroalgae cultivation in effluents, and constructed wetlands (CWs). Lin et al. (2014) highlight that currently, dairy wastewater, including whey, can undergo chemical, thermal, or biochemical transformation processes to convert and or recover particular components with high added values, like protein, lactose, and lactic acid. The organic matter, absorbed in the aquatic macrophyte biomass, can be converted into energy (e.g., biogas, ethanol), used as a fertilizer, for animal

[1] The valorization of waste and effluents, so-called waste recovery, can be understood as a process that transforms the waste or effluent into a co-product with a specific utility, including chemical products, diverse materials, and fuels (Arancon et al. 2013).

feed (Lin et al. 2014), and adsorption of heavy metals and nutrients (Feng et al. 2016).

Luo et al. (2011) studied membranes' use in ultrafiltration and nanofiltration of dairy wastewater treatment and compounds recovering. The authors concluded that the use of membranes allows water reuse and nutrient recovery very efficiently. The authors suggest other processes integrated with the membrane for best performance. For example, produce biogas from the lactose removed by anaerobic digestion. Galanakis et al. (2014) evaluated the use of membranes to extract proteins and whey lactose. The authors concluded the system's viability since the components are used for food purposes due to the high costs.

However, the membrane separation process is not the solution for medium and small industries, as they require investments that, in general, exceed their financial capability. Thus, it is necessary to develop and optimize alternatives according to medium and small dairy industries' purchasing capacity.

An innovative system with different membrane separation processes was proposed, with an annual investment (comprising equipment acquisition and system maintenance) of approximately US$ 3,400,000 for 100 m^3 day^{-1} of effluent (Aydiner et al. 2014). The high operational cost with the energy is another disadvantage for applying membrane cleaning, which must be carried out under pressure due to scale formation (Lin et al. 2014), further worsening dairy industry energy consumption.

Fermentation is a widely studied method for the generation of by-products from waste and effluents, achieving a reduction in BOD and COD rates and the formation of new compounds, such as biofuels (ethanol and biogas) and food additives (lactic acid) (Koutinas et al. 2009). However, in addition to the difficulty of controlling parameters during the process, fermentation is not designed to remove inorganic components like phosphorus. Thus, the environmental problems remain if this wastewater, partially recovered, is discarded in nature. In addition, there is a nutritional waste due to the phosphorus content.

Integrating different procedures would be ideal for carrying out the treatment with recovering and closing the production cycles. An example is the use of Membrane Separation Process associated with the fermentation of the extracted lactose to produce biogas. The biogas would be used as an energy source replacing or reducing firewood use in the boiler. However, small and medium-sized dairy industries generally do not have technical-economic feasibility for implementing the essential technological procedures. Some reasons that hinder such initiatives are limitations of scale production, operating costs, inadequate environmental management, and specialized staff lack to operate certain technologies (Oliveira, Bravo, and Tonial 2012; Prazeres, Carvalho, and Rivas 2012).

Smaller-sized dairy industries can be classified according to the installed capacity for processing the milk received daily. That is: micro up to 5,000 liters; small up to 10,000 liters; medium up to 50,000 liters; large up to 100,000 liters; and exceptional over 100,000 liters (Bahia 2018).

Large companies usually promote the treatment of effluents and the generation of by-products through the implantation of membranes linked to the spray drying equipment - Spray Dryer. In this way, it is possible to efficiently reuse water and nutrients for different purposes (Luo et al. 2011). Processing whey into ricotta or milk beverage is as an option to use whey which is adopted in larger industries. However, the ricotta generates a second whey that still needs treatment. In general, the small dairy industries do not produce milk beverages because it is necessary to set up a new processing line and still compete with established brands. Alves et al. (2019) studied the environmental impacts of using whey to produce dairy beverages, varying the production volume in a small dairy industry. The impacts' contributions are not as relevant from 10% to 40% of the volume range.

2.3.2. Constructed Wetlands in the Treatment of Dairy Wastewater in Small and Medium Companies

Research on wetlands for effluents' treatment is incipient, mainly about the practical application (Matos et al. 2010; Mendonça et al. 2015). Next are shown researches on the use of CWs in the dairy sector. The works cover different purposes since many variables are assessed in each study and directly interfere in the results, such as climate, type of macrophyte, organic and inorganic wastewater load, and hydraulic retention time.

In India, the efficiency of *Hyacinth* sp., *Salvinia molesta*, and *Lemna* sp. for treatment of dairy wastewater as a low-cost option was evaluated. Under laboratory conditions, the results indicated superior COD removals in systems with *Hyacinth* sp and *Salvinia molesta*, compared to *Lemna* sp. *Hyacinth* sp showed marked growth capacity and reductions in the COD values of dairy wastewater even at high initial COD concentrations (2,100 mg L^{-1}) as the authors consider (Bhavsar, Pujari, and Diwan 2010).

The efficiency of six aquatic macrophytes in CWs (*Typha* sp., *Eichhornia* sp., *Salvinia* sp., *Pistia* sp., *Azolla* sp. And *Lemna* sp.) was simulated in the laboratory to assess the wastewater treatment of Indian dairy industries. The results showed *Typha* as the best option for reducing concentrations of pollution in the DW with levels of 99.9%, 83.8%, 70%, 85.5%, and 80% for conductivity, total solids, turbidity, COD, and BOD, respectively, followed by *Eichhornia* sp. and *Salvinia* sp. (Dipu, Kumar, and Thanga 2011).

For Medium and Small Enterprises, mainly in the dairy sector, CWs can represent an attractive system in treating their wastewater, both in the environmental and financial aspects. Lee et al. (2010) highlight the reduced cost for implementing and maintaining wetlands compared to other wastewater treatment methods. Gorra et al. (2014) evaluated a CW performance in an unconventional format to treat wastewater from a medium-sized dairy in different seasons. The results presented that the system had high performance, increasing legal values for wastewater discharge into the environment, and removing pathogens, regardless of the climate (Gorra et al. 2014).

CWs can be applied to treat wastewater from rural areas, including small units, like several properties that benefit dairy products. Mantovi et al. (2003) concluded that the system was efficient in removing biodegradable systems in treating DW in an isolated rural unit. The organic material, achieving a 93.7% reduction in BOD, representing low cost due to the minimal need for material and human resources, being easy to maintain and manage. The authors also recommended that wetlands receive sanitary waste if they do not have a sewage treatment, which would reduce the overall environmental impact of rural property.

Lee et al. (2010) evaluated the efficiency of different wetlands on a pilot scale in treating dairy wastewater from a farm in different seasons. The results showed good performances for the systems studied with greater efficiency of the hybrid system when compared to the integrated system. The hybrid system comprises two CWs disposed of, one in the horizontal and the other in vertical flows, while the integrated system has a steel filter for removing P. On the other hand, the authors indicated the possibility of incorporating the P removal filter to complement the treatment. Change in treatment efficiency was also observed according to the different seasonal effects. The highest rates of removal of nutrients and growth of biomass were found at the end of the summer (Lee et al. 2010).

Matos et al. (2010) evaluated the phytoremediation capacity of tifton 85 grass (*Cynodon* spp.) and elephant grass (*Pennisetum purpureum* Schum.) exposed to different organic loads for the treatment of DW in CW. Ten wetlands were built with the horizontal subsurface flow, of which five were grown with elephant grass and five with 85 tifton grass. The analyzes were carried out for 84 days and with a frequency of 14 days. The organic loads tested were obtained by diluting the DW (with an initial BOD of up to 2,500 mg L-1) with a surface water source. In these conditions, they concluded that tifton grass was more efficient for extracting a more significant amount of N, Na, and equal concentration of P. Therefore, it should be preferred compared to elephant grass and is indicated for reducing pollutants of DW.

The potential for removing N and P from DW in CW was evaluated using the aquatic macrophytes *Typha dominguensis* and *Hedychium coronarium*. The CWs were installed on 200 L drums cut crosswise. The experiment lasted seven months, with analyzes performed every 30 days and a hydraulic retention time of two days for each batch. The efficiency of the CWs decreased throughout the experiment. The authors justified this due to the stabilization of the plant growth, mainly in nitrification or ammonia anaerobic oxidation used by the plant to metabolism its sprouting, growth, and microbiological processes. The authors made no mention of harvesting biomass. It was observed that *Typha dominguensis* and *Hedychium coronarium* could be used to remove N and P from DW. However, *Typha dominguensis* reached the most efficient results, with removal values of 73.4% for total nitrogen and 48.2% for total phosphorus. Finally, the authors conclude that the batch operation is more efficient in removing nutrients (Mendonça et al. 2012).

Mendonça and his collaborators continued researching CWs in DW phytoremediation. This time, they evaluated BOD removal and pH regulation in CW operating in batches and planted with *Typha dominguensis* and the *Hedychium coronarium*. In similar conditions to the previous study, there was no significant difference between the analyzed CWs. The systems showed satisfactory removals of the biodegradable carbonaceous compounds, presenting an average of 90.2% of BOD reductions. The average values for pH were 6, remaining within the limit required by Brazilian legislation for effluents' discharge, CONAMA resolution 430/2011 (Mendonça et al. 2015).

Constructed wetlands using native aquatic macrophytes can be considered an alternative for phytoremediation for medium and small industries (Queiroz et al. 2019; Queiroz et al. 2020). Queiroz et al. (2020) highlighted that the tropical climate contributed to the efficiency in the cycling of N and P by the macrophytes surveyed (e.g., *Cyperus articulates*, *Eichhornia crassipes*, *Eleocharis interstincta* and *Typha domingensis*). The authors also state that the use of CWs in series is a strategy that can increase phytoremediation efficiency, especially when the purpose is to recycle organic and inorganic matter.

2.3.3. Process Integration for Environmental Impact Mitigation and Nutrient Cycling from Aquatic Macrophytes

Alternative treatments for the mitigation of environmental impacts and cycling of energy and nutrients receive different nomenclatures, such as friendly technologies (Eco-friendly) (Zheng et al. 2016), green or blue infrastructure (Derkzen, Teeffelen, and Verburg 2017), industrial ecology (Schiller, Penn, and Basson 2014). Green infrastructure is a technology that seeks to incorporate the principles of natural systems from ecosystems into industrial applications, promoting environmental, social, economic, and health benefits at low costs (Demuzere et al. 2014; DiMuro et al. 2014). Similarly, industrial ecology uses the interface of nature as an example for process integration (Schiller, Penn, and Basson 2014). Demuzere et al. (2014) highlighted that green infrastructure could be composed of natural or built spaces that contribute to ecosystems' resilience and human benefits, such as forests, wetlands, and green roofs.

The integration of processes to close the cycle between the treatment and the reuse of nutrients, water, and energy in the different production stages is essential (Deutz and Ioppolo 2015; Quilliam et al. 2015). Thus, DW treatment with CW can be considered green infrastructure and industrial ecology instead of an alternative treatment of EoP (end-of-pipe). The concern with the treatment only appears after the generation of the wastewater.

The possibility of generating co-products using the aquatic biomass shows the relevance of treating wastewater in CWs with aquatic macrophytes, mainly as an alternative for processes integration in small industries.

The use of aquatic macrophyte biomass ranges from traditional techniques of direct use and more modern processing technologies. Feng et al. (2016) highlighted different ways of harnessing the biomass of *Eichhornia crassipes*, such as fodder and animal feed, mineral compound rich of nutrients (e.g., N, P, K), fermentation for ethanol production, extraction, and recovery of potassium, development of activated carbon, preparation of toxic metal adsorbents, production of biogas and enzymatic hydrolysis for sugar production. However, many studies focus on technical and efficiency issues in the phytoremediation of CWs (Wu et al. 2015). Environmental and economic issues are still developing (Feng et al. 2016) and essential. Such studies must focus on cycling energy and nutrients like the P, which is a finite resource (Quilliam et al. 2015), and with the use of biomass through the integration of processes and reduction of the environmental impact foreseen in industrial ecology (Deutz and Ioppolo 2015; Zhu et al. 2017).

Queiroz et al. (2020) studied the phytoremediation efficiency of four aquatic macrophytes native to northeastern Brazil using in CW: *Cyperus articulatus*, *Eichhornia crassipes*, *Eleocharis interstincta*, and *Typha domingensis*. The phytoremediation efficiency was evaluated by considering the following physicochemical parameters: Chemical Oxygen Demand, Biochemical Oxygen Demand (BOD), Total Solids, Dissolved Solids, Suspended Solids, SedimentableSolids, pH, nitrogen (tN), phosphorus (tP), and weight of aquatic macrophytes and the environmental performance was assessed using the life cycle assessment approach to compare two different scenarios: (S1) animal feed and (S2) biogas/biofertilizer production, for biomass destination of the macrophyte of better efficiency, produced in the treatment.

All four macrophytes reduced BOD by more than 70%, with *Cyperus articulates* being the most efficient in organic load reduction. With the macrophyte *Eichhornia crassipes* arranged in CW in series, the organic load reduction reached average decreases of 75% for tP and 58% for tN.

The S1 treatment was the most favorable alternative for mitigating greenhouse gas emissions and eutrophication. Concurrently, no statistically significant difference was found between the scenarios studied for the impact categories: ozone layer destruction, terrestrial acidification, and cumulative demand for exergy.

CONCLUSION

The wetland system constructed with aquatic macrophytes is a promising technology in treating dairy industry wastewater. It has advantages from the technical and environmental point of view, being a potential alternative to conventional treatments, promoting nutrient cycling, energy recovery, and contributing to reducing local and global environmental impacts associated with the generation of effluents from the food industry.

Several local aquatic plants can be used, and the choice and optimization depend on monitoring the plants' development and maintaining the wastewater treatment. The main criteria are the relationship between development and maintenance with efficiency in removing organic and inorganic loads, seasonality in the development of plants (since there may be a reduction in efficiency during winter), and being a species of easy multiplication and availability in the vicinity of the treatment station. The plant's natural occurrence will also indicate its ability to adapt to the local climate, being one of the essential factors for the development and maintenance of macrophytes during treatment.

Acknowledgments

Conselho Nacional de Desenvolvimento Científico e Tecnológico - CNPq (476725/4572012-7; 313293/2014-7) and Fundação de Amparo à Pesquisa do Estado da Bahia - Fapesb (PAM0017/2014), for the financial support to the development of this research.

References

Alves, Edmar C., Soares, Bruna B., Almeida Neto, José A., Rodrigues, Luciano B. 2019. "Strategies for reducing the environmental impacts of organic mozzarella cheese production." *Journal of Cleaner Production* 223 (20): 226237. https://doi.org/10.1016/j.jclepro.2019.03.006.

ABNT. 2004. "Sistema de Gestão Ambiental- Requisitos Com Orientações Para Uso." *Associação Brasileira de Normas Técnicas-ABNT NBR ISO* 14001:2004. ["Environmental Management System - Requirements with Guidelines for Use." *Brazilian Association of Technical Standards-ABNT NBR ISO* 14001: 2004.]

Ansari, Abid A., and Sarvajeet Singh Gill. 2014. *Eutrophication: Causes, Consequences and Control*. Edited by Abid A. Ansari and Sarvajeet Singh Gill. 2nd ed. Dordrecht: Springer Netherlands. https://doi.org/10.1007/978-94-007-7814-6.

Arancon, Rick Arneil D., Carol Sze Ki Lin, King Ming Chan, Tsz Him Kwan, and Rafael Luque. 2013. "Advances on Waste Valorization: New Horizons for a More Sustainable Society." *Energy Science & Engineering* 1 (2): 53–71. https://doi.org/10.1002/ese3.9.

Aresta, Michele, Angela Dibenedetto, and Grazia Barberio. 2005. "Utilization of Macro-Algae for Enhanced CO2 Fixation and Biofuels Production: Development of a Computing Software for an LCA Study." *Fuel Processing Technology* 86 (14–15): 1679–93. https://doi.org/10.1016/j.fuproc.2005.01.016.

Austin, David C. 2008. *Nitrogen removal system and method for wastewater treatment lagoons.* US7,347,940 B2, issued 2008.

Austin, David C. 2006. *Combined Tidal Vertical Flow and Horizontal Flow Wastewater Treatment System and Method.* 2 (12): 15.

Aydiner, Coskun, Unal Sen, Semra Topcu, Didem Ekinci, Aysegul D. Altinay, Derya Y. Koseoglu-Imer, and Bulent Keskinler. 2014. "Techno-Economic Viability of Innovative Membrane Systems in Water and Mass Recovery from Dairy Wastewater." *Journal of Membrane Science* 458 (May): 66–75. https://doi.org/10.1016/j.memsci.2014.01.058.

Baker, Lawrence A. 2011. "Can Urban P Conservation Help to Prevent the Brown Devolution?" *Chemosphere* 84: 779–84. https://doi.org/10.1016/j.chemosphere.2011.03.026.

Barbosa, A. S.; Florentino, E. R.; Florencio, I. M. 2010. "Utilização Do Soro Como Substrato Para Produção De Aguardente : Estudo Cinético Da Produção De Etanol Aguardiente." *Revista Verde De Agroecologia E Desenvolvimento Sustentável* 5 (1): 7–25. ["Use of Whey as a Substrate for Brandy Production: Kinetic Study of Brandy Ethanol Production." *Green Magazine of Agroecology and Sustainable Development*]

Berlin, Johanna, Ulf Sonesson, and Anne-Marie Tillman. 2007. "A Life Cycle Based Method to Minimise Environmental Impact of Dairy Production through Product Sequencing." *Journal of Cleaner Production* 15 (4): 347–56. https://doi.org/10.1016/j.jclepro.2005.07.019.

Bahia, Conselho Estadual do Meio Ambiente, 2018. Resolução Cepram nº 4.579, de 06 de março de 2018. Altera a Resolução CEPRAM nº 4.327, 31 de outubro de 2013. In: que dispõe sobre as atividades de impacto local de competência dos Municípios, fixa normas gerais de cooperação federativa nas ações administrativas [...], e dá outras providências. *Diário Oficial do Estado da Bahia*, 06 de março de 2018. [Cepram Resolution No. 4,579, of March 6, 2018. Amends CEPRAM Resolution No. 4,327, October 31, 2013. In: which provides for local impact activities within the jurisdiction of the Municipalities, sets general rules for federative cooperation in administrative actions [...], and other measures. *Official Gazette of the State of Bahia*, March 6, 2018]

Bhatia, Misha, and Dinesh Goyal. 2014. "Analyzing Remediation Potential of Wastewater through Wetland Plants: A Review." *Environmental Progress & Sustainable Energy* 33 (1): 9–27. https://doi.org/10.1002/ep.11822.

Bhavsar, Swati R., Vedavati R. Pujari, and V V Diwan. 2010. "Potential of Phytoremediation for Dairy Wastewater Treatment." *Jounal of Mechanical and Civil Engineering*, 16–23.

Brasil. Conselho Nacional do Meio Ambiente - CONAMA. Resolução N° 430 de 13 de Maio de 2011. Dispõe sobre as condições e padrões de lançamento de efluentes. *Diário Oficial da União*, 2011. [Provides for the conditions and standards for the discharge of effluents. *Official Diary of the Union*]

Brião, V. B., and C. R G Tavares. 2007. "Effluent Generation by the Dairy Industry: Preventive Attitudes and Opportunities." *Brazilian Journal of Chemical Engineering* 24 (4): 487–97. https://doi.org/10.1590/S0104-66322007000400003.

Capper, J. L. 2013. "Should We Reject Animal Source Foods to Save the Planet? A Review of the Sustainability of Global Livestock Production." *South African Journal of Animal Sciences* 43 (3): 233–46. https://doi.org/10.4314/sajas.v43i3.1.

Carvalho, Fátima, Ana R Prazeres, and Javier Rivas. 2013. "Cheese Whey Wastewater: Characterization and Treatment." *Science of the Total*

Environment 445–446 (February): 385–96. https://doi.org/10.1016/j.scitotenv.2012.12.038.

Cetesb. 2008. "Guia Técnico Ambiental De Produtos Lácteos." *Guia Técnico Ambiental Da Industria de Produtos Lácteos*, 1–15. https:// doi.org/10.1590/S0103-20032008000200001. [Environmental Technical Guide for Dairy Products. *Environmental Technical Guide for the Dairy Industry*]

Cetesb. 1990. "Companhia de Tecnologia de Saneamento Ambiental." *Nota Sobre Tecnologia de Controle de Produção de Vapor Em Caldeiras a Lenha* São Paulo: 24. [Environmental Sanitation Technology Company. *Note on Steam Production Control Technology in Wood Boilers*]

Cronk, Julie K. 1996. "Constructed Wetlands to Treat Wastewater from Dairy and Swine Operations: A Review." *Agriculture, Ecosystems & Environment* 58 (2–3): 97–114. https://doi.org/10.1016/0167-8809(96)01024-9.

Demuzere, M., K. Orru, O. Heidrich, E. Olazabal, D. Geneletti, H. Orru, A. G. Bhave, N. Mittal, E. Feliu, and M. Faehnle. 2014. "Mitigating and Adapting to Climate Change: Multi-Functional and Multi-Scale Assessment of Green Urban Infrastructure." *Journal of Environmental Management* 146: 107–15. https://doi.org/10.1016/j.jenvman.2014.07.025.

Derkzen, Marthe L., Astrid J A van Teeffelen, and Peter H. Verburg. 2017. "Green Infrastructure for Urban Climate Adaptation: How Do Residents' Views on Climate Impacts and Green Infrastructure Shape Adaptation Preferences?" *Landscape and Urban Planning* 157 (January): 106–30. https://doi.org/10.1016/j.landurbplan.2016.05.027.

Deutz, Pauline, and Giuseppe Ioppolo. 2015. "From Theory to Practice: Enhancing the Potential Policy Impact of Industrial Ecology." *Sustainability* 7 (2): 2259–73. https://doi.org/10.3390/su7022259.

DiMuro, Johnathan L., France M. Guertin, Rich K. Helling, Jessica L. Perkins, and Scanlon Romer. 2014. "A Financial and Environmental Analysis of Constructed Wetlands for Industrial Wastewater

Treatment." *Journal of Industrial Ecology* 18 (5): 631–40. https://doi.org/10.1111/jiec.12129.

Diniz, Célia R, Beatriz S O De Ceballos, José E De L Barbosa, and Annemarie Konig. 2005. "Uso de Macrófitas Aquáticas Como Solução Ecológica Para Melhoria Da Qualidade de Água." *Revista Brasileira de Engenharia Agrícola e Ambiental*, 226–30. ["Use of Aquatic Macrophytes as an Ecological Solution to Improve Water Quality." *Brazilian Journal of Agricultural and Environmental Engineering*]

Dipu, S., Anju A. Kumar, and V. Salom G. Thanga. 2011. "Phytoremediation of Dairy Effluent by Constructed Wetland Technology." *The Environmentalist* 31 (3): 263–78. https://doi.org/10.1007/s10669-011-9331-z.

Djekic, Ilija, Jelena Miocinovic, Igor Tomasevic, Nada Smigic, and Nikola Tomic. 2014. "Environmental Life-Cycle Assessment of Various Dairy Products." *Journal of Cleaner Production* 68 (April): 64–72. https://doi.org/10.1016/j.jclepro.2013.12.054.

Dodds, Walter K, Wes W Bouska, Jeffrey L Eitzmann, Tyler J Pilger, Kristen L Pitts, Alyssa J Riley, Joshua T Schloesser, and Darren J Thornbrugh. 2009. "Eutrophication of U.S. Freshwaters: Analysis of Potential Economic Damages." *Environmental Science & Technology* 43 (1): 12–19. https://doi.org/10.1021/es801217q.

Dunne, E. J., K. R. Reddy, and O. T. Carton, eds. 2005. *Nutrient Management in Agricultural Watersheds*. The Netherlands: Wageningen Academic Publishers. https://doi.org/10.3920/978-90-8686-558-1.

Esteves, Francisco de Assis. 1998. *Fundamentos Da Limnologia*. 2ª. Rio de Janeiro: Interciência. [*Fundamentals Of Limnology*]

Fantin, Valentina, Patrizia Buttol, Roberto Pergreffi, and Paolo Masoni. 2012. "Life Cycle Assessment of Italian High Quality Milk Production. A Comparison with an EPD Study." *Journal of Cleaner Production* 28 (June): 150–59. https://doi.org/10.1016/j.jclepro.2011.10.017.

Feng, Wei, Kai Xiao, Wenbing Zhou, Duanwei Zhu, Yiyong Zhou, Yu Yuan, Naidong Xiao, Xiaoqiong Wan, Yumei Hua, and Jianwei Zhao.

2016. "Analysis of Utilization Technologies for Eichhornia Crassipes Biomass Harvested after Restoration of Wastewater." *Bioresource Technology*. https://doi.org/10.1016/j.biortech.2016.10.047.

Finnegan, William, Jamie Goggins, Eoghan Clifford, and Xinmin Zhan. 2015. "Global Warming Potential Associated with Dairy Products in the Republic of Ireland." *Journal of Cleaner Production*. https://doi.org/10.1016/j.jclepro.2015.08.025.

Galanakis, Charis M., Stathis Chasiotis, George Botsaris, and Vassilis Gekas. 2014. "Separation and Recovery of Proteins and Sugars from Halloumi Cheese Whey." *Food Research International*, 1–7. https://doi.org/10.1016/j.foodres.2014.03.060.

Geraldino, Henrique Cesar Lopes. 2014. *Eletrofloculação Aplicada Ao Tratamento de Efluente Da Indústria de Laticínio*. Universidade Estadual do Sudoeste da Bahia. [*Electrofloculation Applied to Effluent Treatment in the Dairy Industry*]

Golda, A E, G Poyyamoli, and M Nandhivarman. 2014. "Efficacy of Phytoremediation Potential of Aquatic Macrophytes for Its Applicability in Treatment Wetlands : A Review of Developments and Research." *Int. Journal of Water Resources and Environmental Engineering* 6 (10): 267–78. https://doi.org/10.5897/IJWREE2013.0439.

González-García, Sara, Érica G. Castanheira, Ana Cláudia Dias, and Luis Arroja. 2013. "Environmental Performance of a Portuguese Mature Cheese-Making Dairy Mill." *Journal of Cleaner Production* 41: 65–73. https://doi.org/10.1016/j.jclepro.2012.10.010.

Gorra, R., M. Freppaz, E. Zanini, and R. Scalenghe. 2014. "Mountain Dairy Wastewater Treatment with the Use of a 'Irregularly Shaped' Constructed Wetland (Aosta Valley, Italy)." *Ecological Engineering* 73 (December): 176–83. https://doi.org/10.1016/j.ecoleng.2014.09.013.

Headley, T. R., and C. C. Tanner. 2012. "Constructed Wetlands With Floating Emergent Macrophytes: An Innovative Stormwater Treatment Technology." *Critical Reviews in Environmental Science and*

Technology 42 (November 2012): 2261–2310. https://doi.org/10.1080/10643389.2011.574108.

Hobbie, Sarah E. 1992. "Effects of Plant Species on Nutrient Cycling." *Trends in Ecology & Evolution* 7 (10): 336–39. https://doi.org/10.1016/0169-5347(92)90126-V.

Høgaas Eide, Merete. 2002. "Life Cycle Assessment (LCA) of Industrial Milk Production." *The International Journal of Life Cycle Assessment* 7 (2): 115–26. https://doi.org/10.1007/BF02978855.

Hubbard, Robert K. 2010. "Floating Vegetated Mats for Improving Surface Water Quality." In *Emerging Environmental Technologies, Volume II*, 2:211–44. Dordrecht: Springer Netherlands. https://doi.org/10.1007/978-90-481-3352-9_9.

Ibekwe, A Mark, Catherine M Grieve, and Stephen R Lyon. 2003. *Characterization of Microbial Community in Constructed Dairy Wetland.* Pdf 69 (9): 5060–69. https://doi.org/10.1128/AEM.69.9.5060.

Jarvie, Helen P., Colin Neal, and Paul J. A. Withers. 2006. "Sewage-Effluent Phosphorus: A Greater Risk to River Eutrophication than Agricultural Phosphorus?" *Sci Total Environ.* 360 (1–3): 246–53. https://doi.org/10.1016/j.scitotenv.2005.08.038.

Kadlec, Robert, and Scott Wallace. 2008. *Treatment Wetlands, Second Edition. Treatment Wetlands.* 2nd. CRC Press. https://doi.org/10.1201/9781420012514.

Koutinas, Athanasios a, Harris Papapostolou, Dimitra Dimitrellou, Nikolaos Kopsahelis, Eleftheria Katechaki, Argyro Bekatorou, and Loulouda a Bosnea. 2009. "Whey Valorisation: A Complete and Novel Technology Development for Dairy Industry Starter Culture Production." *Bioresource Technology* 100 (15): 3734–39. https://doi.org/10.1016/j.biortech.2009.01.058.

Ladislas, S., C. Gérente, F. Chazarenc, J. Brisson, and Y. Andrès. 2015. "Floating Treatment Wetlands for Heavy Metal Removal in Highway Stormwater Ponds." *Ecological Engineering* 80: 85–91. https://doi.org/10.1016/j.ecoleng.2014.09.115.

Lee, Martin S, Aleksandra Drizo, Donna M Rizzo, Greg Druschel, Nancy Hayden, and Eamon Twohig. 2010. "Evaluating the Efficiency and Temporal Variation of Pilot-Scale Constructed Wetlands and Steel Slag Phosphorus Removing Filters for Treating Dairy Wastewater." *Water Research* 44 (14): 4077–86. https://doi.org/10.1016/j.watres.2010.05.020.

Lin, Carol Sze Ki, Apostolis A. Koutinas, Katerina Stamatelatou, Egid B. Mubofu, Avtar S. Matharu, Nikolaos Kopsahelis, Lucie A. Pfaltzgraff, et al. 2014. "Current and Future Trends in Food Waste Valorization for the Production of Chemicals, Materials and Fuels: A Global Perspective." *Biofuels, Bioproducts and Biorefining* 8 (5): 686–715. https://doi.org/10.1002/bbb.1506.

Lin, Ying Feng, Shuh Ren Jing, Der Yuan Lee, Yih Feng Chang, Yi Ming Chen, and Kai Chung Shih. 2005. "Performance of a Constructed Wetland Treating Intensive Shrimp Aquaculture Wastewater under High Hydraulic Loading Rate." *Environmental Pollution* 134 (3): 411–21. https://doi.org/10.1016/j.envpol.2004.09.015.

Lu, Shibao, Xiaoling Zhang, Jianhua Wang, and Liang Pei. 2016. "Impacts of Different Media on Constructed Wetlands for Rural Household Sewage Treatment." *Journal of Cleaner Production* 127: 325–30. https://doi.org/10.1016/j.jclepro.2016.03.166.

Luo, Jianquan, Luhui Ding, Benkun Qi, Michel Y. Jaffrin, and Yinhua Wan. 2011. "A Two-Stage Ultrafiltration and Nanofiltration Process for Recycling Dairy Wastewater." *Bioresource Technology* 102 (16): 7437–42. https://doi.org/10.1016/j.biortech.2011.05.012.

Ma, Lin, Feng He, Tao Huang, Qiaohong Zhou, Yi Zhang, and Zhenbin Wu. 2016. "Nitrogen and Phosphorus Transformations and Balance in a Pond-Ditch Circulation System for Rural Polluted Water Treatment." *Ecological Engineering* 94: 117–26. https://doi.org/10.1016/j.ecoleng.2016.05.051.

Mantovi, Paolo, Marta Marmiroli, Elena Maestri, Simona Tagliavini, Sergio Piccinini, and Nelson Marmiroli. 2003. "Application of a Horizontal Subsurface Flow Constructed Wetland on Treatment of

Dairy Parlor Wastewater." *Bioresource Technology* 88: 85–94. https://doi.org/10.1016/S0960-8524(02)00291-2.

Martelo, Jorge, and Jaime A Lara Borrero. 2012. "Macrofitas Flotantes En El Tratamiento de Aguas Residuales; Una Revision Del Estado Del Arte." *Ingeniería y Ciencia, Ing. Cienc.* 8 (15): 221–43. https://doi.org/10.5565/rev/athenea.1697. [Floating Macrophytes In *Wastewater Treatment; A Review of the State of the Art. Engineering and Science,* Ing. Cienc]

Matamoros, Víctor, Joan García, and Josep M. Bayona. 2008. "Organic Micropollutant Removal in a Full-Scale Surface Flow Constructed Wetland Fed with Secondary Effluent." *Water Research* 42 (3): 653–60. https://doi.org/10.1016/j.watres.2007.08.016.

Matos, A. T., Sérgio S. Abrahão, Paola a. V. Lo Monaco, Antover P. Sarmento, and Mateus P. Matos. 2010. "Capacidade Extratora de Plantas Em Sistemas Alagados Utilizados No Tratamento de Águas Residuárias de Laticínios." *Revista Brasileira de Engenharia Agrícola e Ambiental* 14 (31): 1311–17. https://doi.org/10.1590/S1415-4366 2010001200009. ["Plant Extraction Capacity in Flooded Systems Used in Dairy Wastewater Treatment." *Brazilian Journal of Agricultural and Environmental Engineering*]

Mendonça, Henrique Vieira, Celso Bandeira de Melo Ribeiro, Alisson Carraro Borges, and Ronaldo Rocha Bastos. 2012. "Remoção de Nitrogênio e Fósforo de Águas Residuáris de Laticínios Por Sistemas Alagados Construídos Operando Em Bateladas." *Revista Ambiente & Agua-An Interdisciplinary Journal of Applied Science* 7 (2): 75–87. https://doi.org/10.4136/1980-993X. ["Removal of Nitrogen and Phosphorus from Dairy Wastewater by Flood Systems Built Operating in Batches." *Revista Ambiente & Agua-An Interdisciplinary Journal of Applied Science*]

Mendonça, Henrique Vieira, Celso Bandeira de Melo Ribeiro, Alisson Carraro Borges, and Ronaldo Rocha Bastos. 2015. "Sistemas Alagados Construídos Em Batelada: Remoção de Demanda Bioquímica de Oxigênio e Regulação de PH No Tratamento de Efluentes de Laticínios." *Ambiente e Água - An Interdisciplinary Journal of Applied*

Science 10 (2): 442–53. https://doi.org/10.4136/1980-993x. ["Batch-Flooded Systems: Biochemical Oxygen Demand Removal and PH Regulation in Dairy Wastewater Treatment."]

Milani, F. X., D. Nutter, and G. Thoma. 2011. "Invited Review: Environmental Impacts of Dairy Processing and Products: A Review." *Jounal Dairy Science* 94 (9): 4243–54. http://dx.doi.org/10.3168/jds.2010-3955.

Oliveira, Débora F. de, Claudia E. C. Bravo, and Ivane B. Tonial. 2012. "Whey of Milk: A Valuable Product." *Revista Do Instituto de Laticínios CâNdido Tostes* 67 (385): 64–71. https://doi.org/10.5935/2238-6416.20120025.

Pelissari, Catiane, Samara Terezinha Decezaro, Pablo Heleno Sezerino, Orlando de Carvalho Júnior, Delmira Beatriz Wolff, and Luiz Sérgio Philippi. 2013. "Wetlands Construídos de Fluxo Vertical Empregado No Tratamento de Efluente de Bovinocultura Leiteira." *Revista Eletrônica de Gestão e Tecnologias Ambientais-GESTA* 1 (2): 223–33. ["Wetlands Constructed of Vertical Flow Used in the Treatment of Dairy Cattle Effluent." *Electronic Journal of Environmental Management and Technologies-GESTA*]

Pompêo, Marcelo. 2008. "Monitoramento e Manejo de Macrófitas Aquáticas." *Oecol. Bras.* 12 (3): 406–24. [Monitoring and Management of Aquatic Macrophytes.]

Prazeres, Ana R, Fátima Carvalho, and Javier Rivas. 2012. "Cheese Whey Management: A Review." *Journal of Environmental Management* 110 (November): 48–68. https://doi.org/10.1016/j.jenvman.2012.05.018.

Queiroz, Rita de Cássia Souza de, Ivon Pinheiro Lôbo, Vinícius de Souza Ribeiro, Luciano Brito; Rodrigues, and José Adolfo de Almeida Neto. 2019. "Assessment of Autochthonous Aquatic Macrophytes with Phytoremediation Potential for Dairy Wastewater Treatment in Floating Constructed Wetlands." *International Journal of Phytoremediation*, November, 1–11. https://doi.org/10.1080/15226514.2019.1686603.

Queiroz, Rita de Cassia Souza de, Henrique Leonardo Maranduba, Murillo Borges Hafner, Luciano Brito Rodrigues, and José Adolfo de Almeida

Neto. 2020. "Life Cycle Thinking Applied to Phytoremediation of Dairy Wastewater Using Aquatic Macrophytes for Treatment and Biomass Production." *Journal of Cleaner Production* 267. https://doi.org/10.1016/j.jclepro.2020.122006.

Quilliam, Richard S., Melanie a. van Niekerk, David R. Chadwick, Paul Cross, Nick Hanley, Davey L. Jones, Andy J.a. Vinten, Nigel Willby, and David M. Oliver. 2015. "Can Macrophyte Harvesting from Eutrophic Water Close the Loop on Nutrient Loss from Agricultural Land?" *Journal of Environmental Management* 152: 210–17. https://doi.org/10.1016/j.jenvman.2015.01.046.

Rivas, Javier, Ana R Prazeres, and Fatima Carvalho. 2011. "Aerobic Biodegradation of Precoagulated Cheese Whey Wastewater." *Journal of Agricultural and Food Chemistry* 59 (6): 2511–17. https://doi.org/10.1021/jf104252w.

Rivas, Javier, Ana R Prazeres, Fatima Carvalho, and Fernando Beltrán. 2010. "Treatment of Cheese Whey Wastewater: Combined Coagulation-Flocculation and Aerobic Biodegradation." *Journal of Agricultural and Food Chemistry* 58 (13): 7871–77. https://doi.org/10.1021/jf100602j.

Rockström, Johan. 2015. *Bounding the Planetary Future : Why We Need a Great Transition*, no. April.

Rockström, Johan, Will Steffen, Kevin Noone, Åsa Persson, F. Stuart Chapin, Eric F. Lambin, Timothy M. Lenton, et al. 2009. "A Safe Operating Space for Humanity." *Nature* 461 (7263): 472–75. https://doi.org/10.1038/461472a.

Saqib, Anam, Muhammad Rizwan Tabbssum, Umer Rashid, Muhammad Ibrahim, Saba Shahid, and Muhammad Aamer Mehmood. 2013. "Marine Macro Algae Ulva : A Potential Feed-Stock for Bio- Ethanol and Biogas Pro." *Asian Journal Agri. Biol.* 1 (3): 155–63.

Saraiva, C. B., R. C. S. Mendonça, A. L. Santos, and D. A. Pereira. 2009. "Water Consumption and Effluent Generation in a Small Dairy Industry." *Rev. Inst. Laticínios Cândido Tostes* 64 (367–368): 10–18.

Schiller, Frank, Alexandra S Penn, and Lauren Basson. 2014. "Analyzing Networks in Industrial Ecology e a Review of Social-Material Network

Analyses." *Journal of Cleaner Production* 76: 1–11. https://doi.org/10.1016/j.jclepro.2014.03.029.

Schirmer, Waldir Nagel, and Gabriela Laila De Oliveira. 2010. "Desodorização De Efluentes Líquidos Tratados Em Leito De Zona De Raízes ('Wetlands')." *Tecno-Lógica*, 11–19. http://online.unisc.br/seer/index.php/tecnologica. ["Deodorization of Liquid Effluents Treated in Bed of Root Zone (' Wetlands ')." *Technological*]

Shaw, S. P., and C. G. Fredine. 1956. *Wetlands of the United States, Their Extent, and Their Value for Waterfowl and Other Wildlife*. Circular 39, U.S. Fish and Wildlife Service, U.S. Department of Interior. Washington. D.C., 1956.

Silva, Luana Viana Costa, Marcus Vinícius Andrade, Kelly Rodrigues, and Glória Marinho. 2013. "Treatment of Synthetic Dairy Wastewater in Batch Reactors Inoculated with *Aspergillus Niger* AN400." *Engenharia Sanitária Ambiental* 18, N 4: 371–80.

Sirivedhin, Tanita, and Kimberly A. Gray. 2006. "Factors Affecting Denitrification Rates in Experimental Wetlands: Field and Laboratory Studies." *Ecological Engineering* 26 (2): 167–81. https://doi.org/10.1016/j.ecoleng.2005.09.001.

Slavov, Aleksandar Kolev. 2017. "DairSlavov, A. K. (2017). *Dairy Wastewaters – General Characteristics and Treatment Possibilities – A Review. Food Technology and Biotechnology*, 55(1). https://doi.org/10.17113/Ftb.55.01.17.4520y Wastewaters – General Characteristics and Treatment Possibilities." *Food Technology and Biotechnology* 55 (1). https://doi.org/10.17113/ftb.55.01.17.4520.

Sperling, Marcos Von. 1996. *Introdução à Qualidade Das Águas e Ao Tratamento de Esgotos*. 2nd ed. Belo Horizonte, 1996.

Sultana, Mar-Yam, Christos Akratos, Dimitrios Vayenas, and Stavros Pavlou. 2015. "Constructed Wetlands in the Treatment of Agro-Industrial Wastewater: A Review." *Hemijska Industrija* 69 (2): 127–42. https://doi.org/10.2298/HEMIND150121018S.

Toma, L, M March, a W Stott, and D J Roberts. 2013. "Environmental Efficiency of Alternative Dairy Systems: A Productive Efficiency

Approach." *Journal of Dairy Science* 96 (11): 7014–31. https://doi.org/10.3168/jds.2013-6911.

Vaccari, David A., Michael Mew, Roland W. Scholz, and Friedrich-Wilhelm Wellmer. 2014. "Exploration: What Reserves and Resources?" In *Sustainable Phosphorus Management*, 129–51. Dordrecht: Springer Netherlands. https://doi.org/10.1007/978-94-007-7250-2_2.

Vymazal, J. 2009. "The Use Constructed Wetlands with Horizontal Sub-Surface Flow for Various Types of Wastewater." *Ecological Engineering* 35 (1): 1–17. https://doi.org/10.1016/j.ecoleng.2008.08.016.

Vymazal, Jan. 2007. "Removal of Nutrients in Various Types of Constructed Wetlands." *Science of the Total Environment* 380 (1–3): 48–65. https://doi.org/10.1016/j.scitotenv.2006.09.014.

Wallace, James M., and Steven I. Safferman. 2014. "Anaerobic Membrane Bioreactors and the Influence of Space Velocity and Biomass Concentration on Methane Production for Liquid Dairy Manure." *Biomass and Bioenergy* 66 (Cmd): 143–50. https://doi.org/10.1016/j.biombioe.2014.02.021.

Wang, Wei, Yi Ding, Jeffrey L. Ullman, Richard F. Ambrose, Yuhui Wang, Xinshan Song, and Zhimiao Zhao. 2016. "Nitrogen Removal Performance in Planted and Unplanted Horizontal Subsurface Flow Constructed Wetlands Treating Different Influent COD/N Ratios." *Environmental Science and Pollution Research* 23 (9): 9012–18. https://doi.org/10.1007/s11356-016-6115-5.

Weaver, John Ernest, and Frederic Edward Clements. 1938. *Plant Ecology*. New York: Macgraw-hill. 1938.

Wu, Haiming, Jian Zhang, Huu Hao Ngo, Wenshan Guo, Zhen Hu, Shuang Liang, Jinlin Fan, and Hai Liu. 2015. "A Review on the Sustainability of Constructed Wetlands for Wastewater Treatment: Design and Operation." *Bioresource Technology* 175: 594–601. https://doi.org/10.1016/j.biortech.2014.10.068.

Wu, Shubiao, Peter Kuschk, Hans Brix, Jan Vymazal, and Renjie Dong. 2014. "Development of Constructed Wetlands in Performance

Intensifications for Wastewater Treatment: A Nitrogen and Organic Matter Targeted Review." *Water Research* 57 (June): 40–55. https://doi.org/10.1016/j.watres.2014.03.020.

Xu, Tengfang, and Joris Flapper. 2009. "Energy Use and Implications for Efficiency Strategies in Global Fluid-Milk Processing Industry." *Energy Policy* 37 (12): 5334–41. https://doi.org/10.1016/j.enpol.2009.07.056.

Yan, Ming-Jia, James Humphreys, and Nicholas M Holden. 2011. "An Evaluation of Life Cycle Assessment of European Milk Production." *Journal of Environmental Management* 92 (3): 372–79. https://doi.org/10.1016/j.jenvman.2010.10.025.

Zehetmeier, M., M. Gandorfer, H. Hoffmann, U. K. Müller, I. J. M. de Boer, and a. Heißenhuber. 2013. "The Impact of Uncertainties on Predicted GHG Emissions of Dairy Cow Production Systems." *Journal of Cleaner Production*, October, 1–9. https://doi.org/10.1016/j.jclepro.2013.09.054.

Zhang, Dong Qing, Soon Keat Tan, Richard M. Gersberg, Junfei Zhu, Sara Sadreddini, and Yifei Li. 2012. "Nutrient Removal in Tropical Subsurface Flow Constructed Wetlands under Batch and Continuous Flow Conditions." *Journal of Environmental Management* 96 (1): 1–6. https://doi.org/10.1016/j.jenvman.2011.10.009.

Zheng, Yucong, Xiaochang Wang, Mawuli Dzakpasu, Yaqian Zhao, Huu Hao Ngo, Wenshan Guo, Yuan Ge, and Jiaqing Xiong. 2016. "Effects of Interspecific Competition on the Growth of Macrophytes and Nutrient Removal in Constructed Wetlands: A Comparative Assessment of Free Water Surface and Horizontal Subsurface Flow Systems." *Bioresource Technology* 207: 134–41. https://doi.org/10.1016/j.biortech.2016.02.008.

Zhu, Fen Meng, Hong Guang Zhu, Wen Yan Shen, and Ting Hua Chen. 2017. "Integrating a Tidal Flow Wetland with Sweet Sorghum for the Treatment of Swine Wastewater and Biomass Production." *Ecological Engineering*. https://doi.org/10.1016/j.ecoleng.2017.01.021.

Biographical Sketches

Rita de Cássia Souza de Queiroz

Affiliation: Federal Institute of Bahia, Jequié, Bahia.

Education: Food Engineer from the State University of Southwest Bahia-UESB (2005), master's degree in Food Science and Technology from the Federal University of Viçosa-UFV (2008) and doctorate in Development and Environment from the State University of Santa Cruz-UESC (2017).

Research and Professional Experience: Professor and researcher at the Federal Institute of Education, Science and Technology of Bahia (IFBA), on the Jequié campus. Member of the research group "Territory, Education and Sustainability (TESSER) -IFBA/Jequié" and, of the research group "Bioenergy and Environment (BioMa) –UESC." Has experience in Food Science and Technology, with an emphasis on Quality Control; in Environmental Sciences, with an emphasis on Industrial Ecology, including perspective of Life Cycle Thinking, Cleaner Production, Circular Economy, Bioremediation, treatment of dairy effluents with aquatic macrophytes, Ecodesign and water resource management with an emphasis on water abstraction from the rain. She currently coordinates and teaches the Lato Sensu Graduate Program in Management and Environmental Education at IFBA/Jequié and is a member of the Municipal Council for the Defense of the Environment (CONDEMA) of Jequié.

Publications from the Last 3 Years:

1. Assessment of autochthonous aquatic macrophytes with phytoremediation potential for dairy wastewater treatment in floating constructed wetlands. *International Journal of Phytoremediation*, v. 22, p. 518-528, 2020.

2. Life cycle thinking applied to phytoremediation of dairy wastewater using aquatic macrophytes for treatment and biomass production. *Journal of Cleaner Production*, v. 267, p. 122006, 2020.

José Adolfo de Almeida Neto

Affiliation: State University of Santa Cruz.

Education: Bachelor (1985) and Master in Agricultural Engineering (1992) from the State University of Campinas (UNICAMP), PhD in Agricultural Engineering from the University of Kassel, Germany (2006).

Research and Professional Experience: Full professor at the Department of Agricultural and Environmental Sciences at the State University of Santa Cruz, acting as a teacher in the undergraduate course in agronomy and geography and as a teacher and coordinator in the graduate course in Development and Environment.

Since the doctorate has researched in Environmental Impact Assessment and Life Cycle Management, specifically in biofuels, energy balance, life cycle inventory analysis, life cycle impact assessment, carbon footprint, and their applications in Industrial Ecology and Clean Production in agricultural, forestry and agro-industrial systems. Member of the National Research Network on Life Cycle Impact Assessment (RAICV) (since 2014), of the Ibero-American Life Cycle Network (RICV) (2016-2029) and alternate member of the Steering Committee of the Brazilian Program for the Assessment of Life Cycle (PBACV) (since 2017).

Publications from the Last 3 Years:

1. Towards a contribution to sustainable management of a dairy supply chain. *Production* (ABEPRO), v. 30, p. 1-13, 2020.

2. Human urine fertiliser in the Brazilian semi-arid: Environmental assessment and water-energy-nutrient nexus. *Science of the Total Environment*, v. 713, p. 136145, 2020.
3. Assessment of autochthonous aquatic macrophytes with phytoremediation potential for dairy wastewater treatment in floating constructed wetlands. *International Journal of Phytoremediation*, v. 22, p. 518-528, 2020.
4. Life cycle thinking applied to phytoremediation of dairy wastewater using aquatic macrophytes for treatment and biomass production. *Journal of Cleaner Production*, v. 267, p. 122006, 2020.
5. Potentials and Limitations of the Bioconversion of Animal Manure Using Fly Larvae. Waste and Biomass Valorization, v. 1, p. 1, 2020.
6. Strategies for reducing the environmental impacts of organic mozzarella cheese production. *Journal of Cleaner Productio*n, v. 223, p. 226-237, 2019.
7. Environmental impact of cheese production. In: Charis M. Galanakis. (Org.). *Environmental Impact of Agro-Food Industry and Food Consumption.* 1ed. Cambridge, EUA: Academic Press, 2021, v. 1, p. 169-187.
8. Contributions of Life Cycle Assessment to the Sustainability of Milk Production. In: Anke Hertz. (Org.). *Dairy Farming: Operations Management, Animal Welfare and Milk Production.* 1ed. New York: Nova Publishers, 2018, v. 1, p. 63-96.

Luciano Brito Rodrigues

Affiliation: State University of Southwest Bahia.

Education: Bachelor (1998) and Master (2001) in Mechanical Engineering from the Federal University of Paraíba and doctorate (2008) in Mechanical Engineering from the Federal University of Minas Gerais, with sandwich period (2007-2008) at the Instituto Superior Técnico of Lisbon, Portugal. Post-doctorate (2014-2015) at the State University of Santa Cruz

- UESC, in Development and Environment, with an emphasis on Environmental Management.

Research and Professional Experience: Full Professor at the State University of the Southwest of Bahia - UESB, Food Engineering and Environmental Engineering in undergraduate courses. Is permanent professor in the graduate programs in Food Engineering and Science (PPGEAL) and in Environmental Sciences (PPGCA), both at UESB, and also a permanent professor in the Graduate Program in Development and Environment (PPGDMA) at UESC.

Works in the area of Engineering and Environmental Sciences, in the research lines (i) Environmental Sustainability of Industrial Products and Processes; (ii) Development of Technologies and Solutions for Sustainable Development. The working methods include Life Cycle Management, Life Cycle Assessment, Cleaner Production, Circular Economy, Integrated Management, Environmental Labeling. He is a member of the Advances in Cleaner Production Network (ACPN) and the Life Cycle Impact Assessment Research Network (RAICV). Head of the Materials and Environment Research Group - GEMMA, and Coordinator of the Materials Testing Laboratory - LabEM.

Publications from the Last 3 Years:

1. Towards a contribution to sustainable management of a dairy supply chain. *Production* (ABEPRO), v. 30, p. 1-13, 2020.
2. Assessment of autochthonous aquatic macrophytes with phytoremediation potential for dairy wastewater treatment in floating constructed wetlands. *International Journal of Phytoremediation*, v. 22, p. 518-528, 2020.
3. Life cycle thinking applied to phytoremediation of dairy wastewater using aquatic macrophytes for treatment and biomass production. *Journal of Cleaner Production*, v. 267, p. 122006, 2020.

4. Strategies for reducing the environmental impacts of organic mozzarella cheese production. *Journal of Cleaner Production*, v. 223, p. 226-237, 2019.
5. Effect of handling and feeding strategies in the environmental performance of buffalo milk in Northeastern Brazil. The International *Journal of Life Cycle Assessment*, v. 24, p. 1129-1138, 2019.
6. Environmental impact of cheese production. In: Charis M. Galanakis. (Org.). *Environmental Impact of Agro-Food Industry and Food Consumption.* 1ed. Cambridge, EUA: Academic Press, 2021, v. 1, p. 169-187.
7. Contributions of Life Cycle Assessment to the Sustainability of Milk Production. In: Anke Hertz. (Org.). *Dairy Farming: Operations Management, Animal Welfare and Milk Production.* 1ed. New York: Nova Publishers, 2018, v. 1, p. 63-96.

In: The Future of Phytoremediation
Editor: James C. Flores
ISBN: 978-1-53619-625-2

Chapter 5

EFFECT OF PLANT SPECIES ON MICROBIAL ACTIVITY OF A SUBSTRATE USED IN PHYTOREMEDIATION

Maria G. Alifragki[1,*], PhD,
Athina K. Pavlatou-Ve[2], PhD,
and Michail Z. Orfanoudakis[1], PhD
[1]Department of Forestry and Management of the Environment and Natural Resources, Democritus University of Thrace, Orestiada, Greece
[2]Department of Hydraulics, Soil Science and Agricultural Engineering, School of Agriculture, Aristotle University of Thessaloniki, Thessaloniki, Greece

ABSTRACT

Plant species effect on microbial activity of a limestone quarry with pyrolusite's byproducts mix as cover was investigated. For the

* Corresponding Author's E-mail: mariaalifragki@yahoo.com.

remediation project three plant species: pine, cypress, and broom were planted at 9 randomized plots. Microbial biomass (C_{mic}), basal respiration (BR), alkaline phosphatase activity (ALP), along with microbial quotient (C_{mic}/C_{org} ratio), metabolic quotient (qCO_2), and mineralization quotient (qM) were determined for a 3.5 years period from samples obtained from the rhizosphere and they used as indicators of substrate biological activity. These showed seasonal variation depending on substrate's moisture content. During the first year of phytoremediation microbial activity under broom was significantly higher than under the other two plant species (C_{mic}=225.81 μg C g^{-1}, C_{mic}/C_{org} ratio=6.04%, qCO_2=0.47 μg CO_2-C $μg^{-1}$ C_{mic} d^{-1} and qM=0.12 Σμg CO_2-C $μg^{-1}$ C_{org}). Two years after vegetation establishment, the pine treatment observed significantly higher microbial activity (C_{mic}=366.24 μg C g^{-1}, BR=120.8 μg CO_2-C g^{-1} d^{-1}, ALP=339.13 μg *p*-nitrophenol g^{-1} h^{-1}, C_{mic}/C_{org} ratio=5.85%, qCO_2=0.45 μg CO_2-C $μg^{-1}$ C_{mic} d^{-1} and qM=0.10 Σμg CO_2-C $μg^{-1}$ C_{org}) than broom and cypress treatments. However, at the same year but after a drought period with moisture as low as 10%, microbial activity under cypress plants was significantly higher than in the pine or broom rhizosphere (C_{mic}=43.49 μg C g^{-1}, BR=106.4 μg CO_2-C g^{-1} d^{-1}, C_{mic}/C_{org} ratio=1.07%, qCO_2=3.21 μg CO_2-C $μg^{-1}$ C_{mic} d^{-1} and qM=0.14 Σμg CO_2-C $μg^{-1}$ C_{org}). Soil moisture, was in positive correlation with C_{mic}, ALP (r=0.31 and r=0.70, $p < 0.001$, respectively) and C_{mic}/C_{org} ratio (r=0.15, $p < 0.05$) while it was negative with BR (r=-0.19, $p < 0.01$), qCO_2 and qM (r=-0.29 and r=-0.38, $p < 0.001$, respectively).

Keywords: phytoremediation, microbial activity, plant species

INTRODUCTION

Mining and quarrying works are leaving open scars over the land and phytoremediation is an effective process to minimize the optical disturbance caused by them. Phytoremediation is considered to be successful when it is permanent, aesthetically pleasing and relatively affordable (Tordoff et al. 2000). The quest to find soil with the appropriate quality to cover the land of disturbed sites constitutes is a challenging task in any effort of restoration (Luna et al. 2016). To confront this problem, soil is imported from elsewhere (Land Use Consultants 1992) or is created on site from indigenous materials. These types of substrates though do not always acquire the desired characteristics for the development of vegetation (Legwaila et al. 2015). The land cover mix material is assessed i) directly through the monitoring of its properties and ii) indirectly through its assessment along the course of vegetation establishment in the area to be restored.

Soil quality is determined from its physical and chemical properties which are, however, changing extremely slowly and it takes many years in order to produce quantifiable changes (Pascual et al. 2000). Microbial community determination is one way to tackle this problem (Harris 2003). Soil microorganisms are important to soil development and conservation and they show direct reaction to the environmental changes. Soil microbial biomass (C_{mic}), soil respiration (BR) and enzyme activity in particular, can be used as indicators of quality changes (Anderson and Domsch 1990; Brookes 1995; Fernandes et al. 2005; Giller et al. 1998; Wardle and Ghani 1995) and have been suggested as a way of assessment for its restoration efforts (Graham and Haynes 2004).

Microbial quotient (C_{mic}/C_{org} ratio) (Insam and Domsch 1988), the metabolic quotient (qCO_2) (Anderson and Domsch 1993; Saviozzi et al. 2001) and mineralization quotient (qM) (Moscatelli et al. 2005) have been used for the same purpose as they are indicators of the substrate's quality. They are also used as ecosystem restoration and soil disturbance indicators and are considered more reliable than their individual parts.

The aforesaid soil biological properties as mentioned several times in the literature, could be affected by soil moisture (Criquet 2002; Speir and Cowling 1991; Tang et al. 2006) and they show a seasonal variation (Baldrian et al. 2008; Kamala and Sunada 2014; Nazaries et al. 2015). Moreover, the composition of soil microbial community is affected by vegetation (Roldán et al. 1994).

The soil substrate of such sites starts to show changes upon its properties. The vegetation starts to affect the substrate, as soon as the plant roots grown inside it. The soil surrounding the root system differ drastically from the rest of the soil (Bertin et al. 2003). The physical, chemical and biological differences are caused by the plethora of chemical compounds that are released by the plant in the rhizosphere during its development. Vegetation can also affect the soil's microenvironment

(Raich and Tufekcioglu 2000), the density and the development of roots (DiMichele et al. 2006) and the quantity and quality of plant residues (Raich and Tufekcioglu 2000). Furthermore, the compounds that are secreted from the roots differ depending on the species and age of the plant and the environmental and soil conditions (Pandey and Palni 2007; Sinha et al. 2009).

Plants are carbon and other nutrients source for soil microorganisms and it has been confirmed that soil microbial and enzyme activity increase with vegetation establishment (eg George et al. 2002; Stroo and Jencks 1982). The effect of a different plant species on biological activity under comparable soil and climatic conditions has been studied for several years (Bauhus et al. 1998; Masto et al. 2006; Sparling et al. 1994).

The development of microbial activity in substrate for quarry restoration was assessed under different plant species in relation to the soil moisture content. To test this, we used a limestone quarry cover mixture consisting mainly of pyrolusite's by-products, where planted pine, cypress, and broom plant. Microbial biomass (C_{mic}), basal respiration (BR), alkaline phosphatase activity (ALP), along with microbial quotient (C_{mic}/C_{org}), metabolic quotient (qCO_2), and mineralization quotient (qM) were determined on substrate samples taken from the limestone quarry area during 3.5 year of phytoremediation and were used as rehabilitation indicators.

MATERIALS AND METHODS

Study Area

The experiment was conducted at a limestone quarry in Thessaloniki (North of Greece, 40°44'96''N and 23°04'13''E). The climate of this area is characterized as sub-Mediterranean. A part of the quarry (about 0.3 ha) was covered with a mixture used as substrate for quarry restoration. The substrate mixture consisted of 70% by-products from pyrolusite (MnO_2) electrolytic process, 10% sand (to improve the physical properties), 10%

soil (to supply the mixture with microorganisms) and 10% rice husk (to improve the mixture properties).

Sampling Method

The experiment started in May 2011, when the limestone quarry was covered with the substrate mixture and three plant species: pine (*Pinus brutia*), cypress (*Cupressus sempervirens*) and broom (*Spartium junceum*) were planted with a completely randomized design with three replications. The experiment had 9 plots (10 × 10 m) and each plot was planted with 36 plants. At this time, five random substrate samples of the mixture were collected and used as control. The properties of substrate were: 8.14 pH, 0.68% organic carbon (C_{org}) content, 0.05% N, 85.84 μg C g^{-1} microbial biomass (C_{mic}), 78.80 μg CO_2-C g^{-1} d^{-1} basal respiration and 108.21 μg *p*-nitrophenol g^{-1} h^{-1} alkaline phosphatase activity (ALP). Biological parameters including microbial biomass C, basal respiration and phosphatase activity were monitored for 3.5 years. In March 2012 (1st sampling), September 2012 (2nd sampling), April 2013 (3rd sampling), October 2013 (4th sampling) and October 2014 (5th sampling) substrate samples were collected at 0-10 cm in the rhizosphere of plants, from each plot. The samples were sieved to ≤ 2 mm and divided into two parts. The first part was air-dried and used for chemical analyses, while the other was stored at 4°C in plastic bags and used for microbial analyses.

Analytical Methods

The moisture content of substrate samples was determined as the mass of water in the sample expressed as a percentage of the dry mass, heating at 105°C for 24 h. The content of total organic carbon (C_{org}) was estimated following the method reported by Nelson and Sommers (1982).

Microbial biomass C (C_{mic}) was estimated following the chloroform – fumigation extraction method (Rowell 1994). Two portion of moist substrate samples were weighed (25 g), the first one was fumigated with ethanol-free $CHCl_3$ for 24 h and then extracted with 100 ml of 2 M KCl by shaking for 30 min and filtered (Whatman no. 42). The other non-fumigated portions were extracted similarly. All results are expressed as a means of triplicate determinations in μg C g^{-1}. For the different plant species and for each substrate sample, results were calculated as the rate of microbial biomass variation compared to the control using the equation: $\Delta C_{mic} = (C_{mic1}-C_{mic0})/C_{mic1}$, where ΔC_{mic} is the microbial biomass rate, C_{mic1} is the mean microbial biomass C of each sample and C_{mic0} is the mean microbial biomass C of control.

For measuring microbial respiration air-dried substrate samples (50 g) were adjusted to 45% of maximum water holding capacity and were incubated at 22°C in gas-tight jars. Respiration was determined by quantifying the CO_2 released in the process of microbial respiration between one (basal respiration -BR) and 15 days of incubation. The released CO_2 was trapped in 10 ml 0.3 M NaOH and determined by titration of the excess of hydroxide with 0.2 M HCl after precipitating the carbonate with $BaCl_2$ solution, in the presence of phenolphthalein as indicator (Rowell 1994). The CO_2 evolved from basal respiration was expresses as μg CO_2-C g^{-1} d^{-1}. For the different plant species and for each substrate sample, results were calculated as the rate of basal respiration variation compared to the control using the equation: $\Delta BR = (BR_1-BR_0)/BR_1$, where ΔBR is the basal respiration rate, BR_1 is the mean respiration of each sample and BR_0 is the mean respiration of control.

Alkaline phosphatase activity (ALP) was measured using *p*-nitrophenol as a substrate and pH 11 buffer (Tabatabai 1994). An amount of 0.25 ml toluene, 4 ml modified universal buffer (MUB) (pH 11) and 1 ml *p*-nitrophenyl phosphate solution were added in 1 g of moist substrate. The mixture was held at 37°C for 1 h. After the incubation, 1 ml $CaCl_2$ (0.5 M) and 4 ml NaOH (0.5 M) were added. Then samples were swirled and filtered (Whatman no. 2v). The optical density of the filtrate was measured by a u.v.-visible spectrophotometer set at 400 nm. Results are expressed as μg *p*-nitrophenol g^{-1} h^{-1} at 37°C. For the different plant species and for each substrate sample, results were calculated as the rate of variation of alkaline phosphatase activity compared to the control using the equation: $\Delta ALP = (ALP_1 - ALP_0)/ALP_1$, where ΔALP is the alkaline phosphatase activity rate, ALP_1 is the mean alkaline phosphatase activity of each sample and ALP_0 is the mean alkaline phosphatase activity of control.

Microbial indices could be utilized as indicators of substrate biological properties and were calculated as follows. Microbial quotient or C_{mic}/C_{org} ratio was calculated as reported by Anderson and Domsch (1989): $C_{mic}/C_{org} = \mu g\ C_{mic}\ \mu g\ C_{org}^{-1}$, where C_{mic} is microbial biomass C content and C_{org} is total organic C content. Metabolic quotient (qCO_2) was calculated as: $qCO_2 = \mu g\ CO_2\text{-}C_{basal}\ h^{-1}\ \mu g^{-1}\ C_{mic}^{-1}$ (Dilly and Munch 1998), where $CO_2\text{-}C_{basal}$ is CO_2-C released from the basal respiration and C_{mic} is microbial biomass C content. Mineralization quotient (qM) was calculated as: $qM = \mu g\ CO_2\text{-}C_{cumulative}\ \mu g\ C_{org}^{-1}$ (Pinzari et al. 1999), where $CO_2\text{-}C_{cumulative}$ is CO_2-C released from respiration of 15 days incubation and C_{org} is total organic C content.

Statistical Analyses

The data analyses were performed in triplicate and the mean values were used in further calculations. The data values presented in the text are means values of nine replicates per treatment. Differences in microbial properties moisture and organic C content and microbial indices between treatments were tested by one-way analysis of variance (ANOVA). The Fisher least significant difference (LSD) test for multiple comparisons was run if significant differences were found ($p < 0.05$). Relationships between the measured chemical and microbial properties of the substrate mixture samples were studied using Pearson correlations. The statistical calculations were performed with Statgraphics Centurion XV.

RESULTS

Moisture Content

The moisture content of substrate under pine, cypress and broom plants ranged from 7.00 at 4th sampling to 33.29% at control (Figure 1). The values of moisture content during the 4th sampling were generally low (<10%) in all treatments, due to prolonged drought period (data not shown). At 3rd and 4th sampling the values of substrate's moisture under pines (15.40 and 8.14%, respectively) were significantly higher than cypress (13.11 and 7.08%, respectively) and broom (13.41 and 7.00%, respectively) plants.

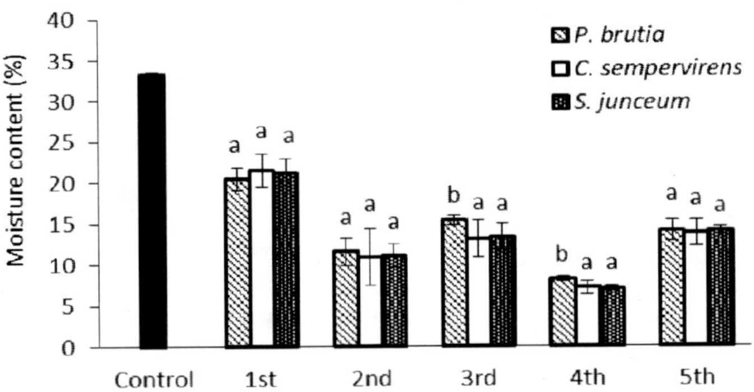

Figure 1. Mean values (n = 9) of moisture content (%) of the substrate under *Pinus brutia*, *Cupressus sempervirens* and *Spartium junceum*, during the experimental time (control, 1st to 5th sampling). Error bars indicate standard errors. Columns sharing the same latter do not differ significantly ($p < 0.05$).

Organic Carbon

Organic carbon (C_{org}) of the substrate under pine, cypress and broom plants varied significantly between samplings and the lowest C_{org} values were at the 2nd sampling for all plant treatments (Figure 2). C_{org} content under cypress plants (0.52%) was significantly higher than broom treatment (0.37%) during the 1st sampling (Table 1). At 3rd and 4th sampling pine treatment (0.62 and 0.66%, respectively) had significantly higher C_{org} content than broom (0.53%) and cypress (0.43%) treatment respectively (Table 1). At the end of experimental time (5th sampling) the C_{org} content under cypress plants (0.77%) was significantly higher than under pine (0.47%) and broom (0.44%) plants (Table 1).

Table 1. Organic C (C_{org}), microbial biomass C (C_{mic}), basal respiration (BR), microbial quotient (C_{mic}/C_{org}), metabolic quotient (qCO_2), mineralization quotient (qM) and alkaline phosphatase activity (ALP) of the substrate vegetated with *Pinus brutia*, *Cupressus sempervirens* and *Spartium junceum*, during the experimental time

	C_{org}	C_{mic}	BR	C_{mic}/C_{org}	qCO_2	qM	ALP
	%	μg C g^{-1}	μg CO_2-C g^{-1} d^{-1}	%	μg CO_2-C μg^{-1} C_{mic} d^{-1}	Σμg CO_2-C μg^{-1} C_{org}	μg *p*-nitrophenol g^{-1} h^{-1}
1st sampling							
P. brutia	0.45 ab	75.35a	84.0a	1.62a	1.24ab	0.09a	387.16a
C. sempervirens	0.52b	99.42a	80.2a	1.77a	4.09b	0.09a	329.08a
S. junceum	0.37 a	225.81b	87.4a	6.04b	0.47a	0.12b	381.06a
2nd sampling							
P. brutia	0.27 a	112.48 b	94.8 a	4.44 ab	1.10 a	0.16 a	168.52 a
C. sempervirens	0.22 a	59.13 a	91.2 a	2.44 a	4.42 b	0.20 a	199.23 a
S. junceum	0.27 a	123.72 b	98.8 a	5.01 b	0.81 a	0.17 a	191.82 a
3rd sampling							
P. brutia	0.62 b	366.24 b	120.8 b	5.85 b	0.45 a	0.10 b	339.13 b
C. sempervirens	0.54 ab	106.38 a	103.2 a	2.01 a	1.08 b	0.09 a	244.25 a
S. junceum	0.53 a	82.72 a	96.8 a	1.58 a	1.58 b	0.08 a	250.83 a
4th sampling							
P. brutia	0,66 b	15,16 a	98,0 ab	0,24 a	8,36b	0,09 a	148,36 a
C. sempervirens	0,43 a	43,49 b	106,4 b	1,07 b	3,21 a	0,14 b	123,82 a
S. junceum	0,53 ab	18,25 a	87,2 a	0,40 a	6,86 b	0,09 a	111,59 a
5th sampling							
P. brutia	0,47 a	140,33 a	93,2 a	3,06 a	0,98 b	0,10 a	198,92 a
C. sempervirens	0,77 b	214,94 a	112,8 b	3,03 a	0,80 ab	0,09 a	176,89 a
S. junceum	0,44 a	156,26 a	82,0 a	4,24 a	0,53 a	0,11 a	173,88 a

Values in columns and within the same sampling sharing the same letter do not differ significantly at $p<0.05$.

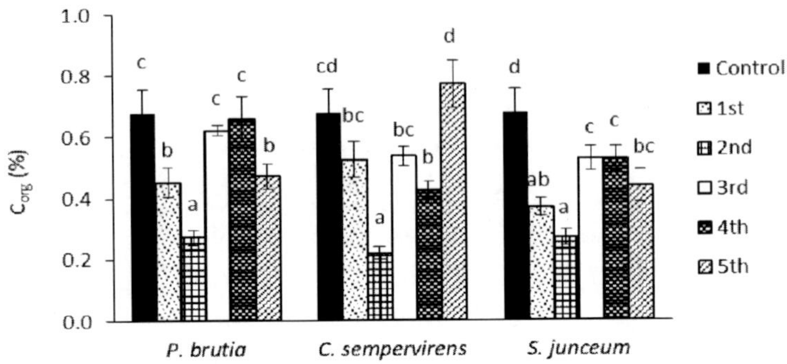

Figure 2. Mean values (n = 9) of the organic C (C_{org}) of the substrate under *Pinus brutia*, *Cupressus sempervirens* and *Spartium junceum*, during the experimental time (control, 1st to 5th sampling). Error bars indicate standard errors. Columns sharing the same latter do not differ significantly ($p < 0.05$).

Microbial Biomass C

Microbial biomass C rate (ΔC_{mic}) values varied among samplings within the range of -4.66 to 0.77 and were almost similar (around zero) at 1st, 2nd, 3rd and 5th sampling (Figure 3A). However, ΔC_{mic} had the lowest level during the 4th sampling at all treatments, when moisture content was low (Figure 3A).

Microbial biomass C (C_{mic}) content of substrate under broom plants (225.81 μg C g^{-1}) was significantly higher than under pine (75.35 μg C g^{-1}) and cypress (99.42 μg C g^{-1}) plants at 1st sampling (Table 1). During the 2nd sampling lower C_{mic} values observed at cypress treatment (59.13 μg C g^{-1}) than at pine and broom treatments (112.48 and 123.72 μg C g^{-1}, respectively) (Table 1). At 3rd sampling C_{mic} values at pine treatment reached to 366.24 μg C g^{-1} and was significantly higher in comparison to cypress and broom treatments (106.38 and 82.72 μg C g^{-1}, respectively) (Table 1). Furthermore, at 4th sampling the highest C_{mic} content was determined under cypress plants (43.49 μg C g^{-1}) (Table 1).

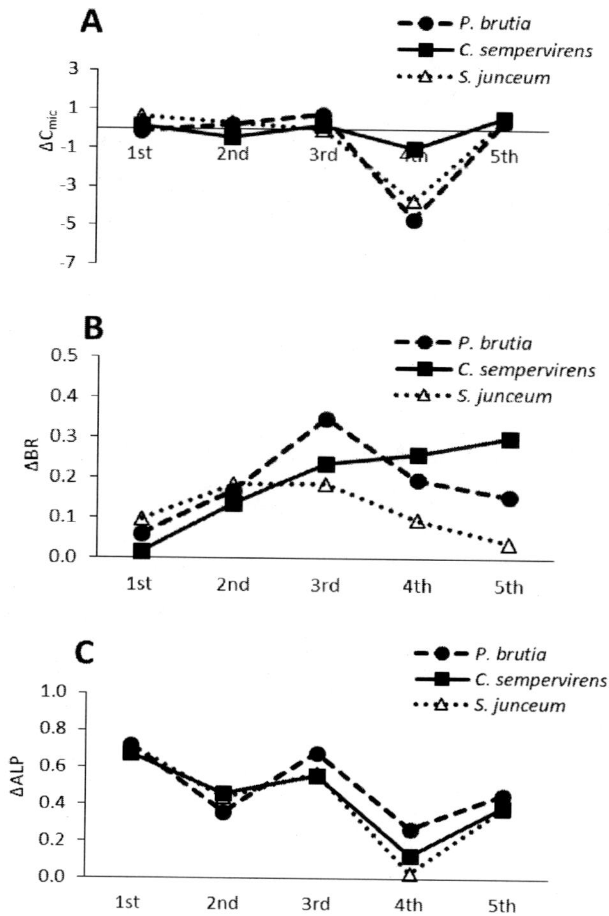

Figure 3. Variation of (A) microbial biomass (ΔC_{mic}), (B) basal respiration (ΔBR) and (C) alkaline phosphatase activity (ΔALP) rate in relation to the control for the substrate under *Pinus brutia*, *Cupressus sempervirens* and *Spartium junceum*, during the experimental time (1st to 5th sampling).

Basal Respiration

Basal respiration rate (ΔBR) of substrate under cypress plants had increasing trend during the study period (Figure 3B). ΔBR of substrate under pine and broom plants increased from the 1st to 3rd sampling and then decreased until the end of experimental time (Figure 3B). Basal respiration (BR) of substrate during the 3rd sampling was significantly higher under pine plants (120.8 μg CO_2-C g^{-1} d^{-1}) than under cypress and broom plants (103.2 and 96.8 μg CO_2-C g^{-1} d^{-1}, respectively) (Table 1). At 4th and 5th samplings the highest BR values were determined under cypress plants (106.4 and 112.8 μg CO_2-C g^{-1} d^{-1}, respectively) (Table 1).

Alkaline Phosphatase Activity

Alkaline phosphatase activity rate (ΔALP) showed seasonal variation during the study period (Figure 3C). ΔALP values varied among samplings within the range of 0.03 to 0.72. ΔALP values were higher in springs than in autumn at all treatments and the minimum rates occurred at 4th sampling, when moisture content was really low (<10%) (Figure 3C). Differences between treatments were observed only at 3rd sampling, when ALP activity under pine plants (339.13 μg *p*-nitrophenol g^{-1} h^{-1}) was significantly higher in comparison to cypress and broom treatments (244.25 and 250.83 μg *p*-nitrophenol g^{-1} h^{-1}, respectively) (Table 1).

Microbial Quotient

Microbial quotient (C_{mic}/C_{org}) values ranged from 0.24 to 6.04% and varied among samplings within the same plant treatment (Figure 4A). The lowest C_{mic}/C_{org} values were calculated during the 4th sampling for all treatments (Figure 4A). At 1st and 2nd sampling significantly higher C_{mic}/C_{org} values between the treatments were under broom plants (6.04 and 5.01%, respectively) (Table 1). At the 3rd sampling C_{mic}/C_{org} value at pine treatment (5.85%) was significantly higher in comparison to cypress and broom treatments (2.01 and 1.58%, respectively) (Table 1). Finally, at 4th sampling the highest C_{mic}/C_{org} value was determined under cypress (1.07%) plants (Table 1).

Metabolic Quotient

Metabolic quotient (qCO_2) values of substrate ranged between 0.45 and 8.36 μg CO_2-C μg^{-1} C_{mic} d^{-1}. For pine and broom treatments the significantly highest qCO_2 values were calculated at the 4th sampling (Figure 4B). At cypress treatment qCO_2 values varied among samplings during the study period (Figure 4B).

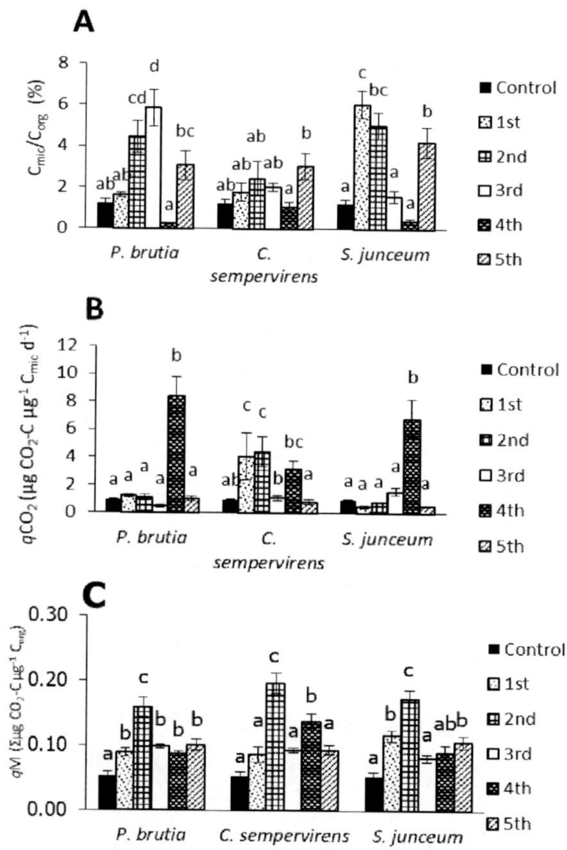

Figure 4. Mean values (n = 9) of (A) microbial quotient (C_{mic}/C_{org}), (B) metabolic quotient (qCO_2) and (C) mineralization quotient (qM) of the substrate under *Pinus brutia*, *Cupressus sempervirens* and *Spartium junceum*, during the experimental time (control, 1st to 5th sampling). Error bars indicate standard errors. Columns sharing the same latter do not differ significantly ($p < 0.05$).

Data analysis showed that 1st sampling the lowest qCO_2 value was measured in broom treatment (0.47 μg CO_2-C $μg^{-1}$ C_{mic} d^{-1}) and the highest in cypress treatment (4.09 μg CO_2-C $μg^{-1}$ C_{mic} d^{-1}) (Table 1). Furthermore, at the 2nd sampling qCO_2 value was significantly higher at cypress treatment (4.42 μg CO_2-C $μg^{-1}$ C_{mic} d^{-1}) in comparison to pine and broom treatments (1.10 and 0.81 μg CO_2-C $μg^{-1}$ C_{mic} d^{-1}, respectively) (Table 1). The lowest qCO_2 values were determined under pine plants (0.45 μg CO_2-C $μg^{-1}$ C_{mic} d^{-1}) at the 3rd sampling and under cypress plants (3.21 μg CO_2-C $μg^{-1}$ C_{mic} d^{-1}) at 4th sampling (Table 1). Furthermore, during the final (5th) sampling qCO_2 was significantly higher under pine plants (0.98 μg CO_2-C $μg^{-1}$ C_{mic} d^{-1}) than under broom plants (0.53 μg CO_2-C $μg^{-1}$ C_{mic} d^{-1}) (Table 1).

Mineralization Quotient

Mineralization quotient (qM) values ranged from 0.05 to 0.20 Σμg CO_2-C $μg^{-1}$ C_{org} and varied among samplings (Figure 4C). qM of substrate increased significantly until the 2nd sampling in relation to the control (Figure 4C), while during the same period C_{org} content was decreased (Figure 2). qM of substrate at 1st sampling was significantly higher under broom plants (0.12 Σμg CO_2-C $μg^{-1}$ C_{org}) than under pine and cypress plants (0.09 and 0.09 Σμg CO_2-C $μg^{-1}$ C_{org}, respectively) (Table 1). During the 3rd sampling qM values under pine plants (0.10 Σμg CO_2-C $μg^{-1}$ C_{org}) were significantly higher in comparison to cypress and broom treatments (0.09 and 0.08 Σμg CO_2-C $μg^{-1}$ C_{org}, respectively) (Table 1). Also, at 4th sampling qM was significantly higher at cypress treatment (0.14 Σμg CO_2-C $μg^{-1}$ C_{org}) than under pine and broom plants (0.09 and 0.09 Σμg CO_2-C $μg^{-1}$ C_{org}, respectively) (Table 1).

Correlation between Substrate Mixture Parameters

Correlation between biological parameters (C_{mic}, BR, C_{mic}/C_{org}, qCO_2, qM, ALP), organic carbon content (C_{org}) and substrate moisture for all treatments and all samplings are presented at Table 2. The substrate moisture was significantly correlated with all other parameters (C_{org}, C_{mic}, BR, C_{mic}/C_{org}, qCO_2, qM, ALP). The C_{mic} and BR were correlated with each other (r = 0.45, p < 0.001). C_{mic} and BR were strongly positively related to the content of C_{org} (r = 0.27 and 0.57, p < 0.001, respectively) and to C_{mic}/C_{org} ratio (r = 0.79, p < 0.001 and 0.14, p < 0.05, respectively).

Table 2. Person's correlation coefficients (r) between vegetated substrate parameters for all treatments and all years (n=270)

	Moisture	C_{org}	C_{mic}	BR	C_{mic}/C_{org}	qCO_2	qM	ALP
Moisture	1							
C_{org}	0.22***	1						
C_{mic}	0.31***	0.27***	1					
BR	-0.19**	0.57***	0.45***	1				
C_{mic}/C_{org}	0.15*	-0.22***	0.79***	0.14*	1			
qCO_2	-0.29***	ns	-0.57***	-0.17**	-0.57***	1		
qM	-0.38***	-0.77***	ns	ns	0.41***	ns	1	
ALP	0.70***	0.29***	0.42***	ns	0.22***	-0.26***	-0.31***	1

C_{org}: organic carbon, C_{mic}: microbial biomass, BR: basal respiration, C_{mic}/C_{org}: microbial quotient, qCO_2: metabolic quotient, qM: mineralization quotient, ALP: alkaline phosphatase activity.

ns, not significant, * significant at p<0.05, ** significant at p<0.01, *** significant at p<0.001.

Also, there was negative correlation among C_{mic}/C_{org} ratio and qCO_2 (r = -0.57, p < 0.001) and strong positive correlation between C_{mic}/C_{org} and qM and ALP (r = 0.41 and r = 0.22, p < 0.001, respectively). The qCO_2 was negatively correlated with C_{mic} (r = -0.57, p < 0.001), with BR (r = -0.17, p < 0.01) and with ALP (r = -0.26, p < 0.001). Furthermore, ALP was positively and significantly affected by C_{org} (r = 0.29, p < 0.001) and C_{mic} (r = 0.42, p < 0.001) and was related negatively with qM (r = -0.31, p < 0.001). Finally, there was strong negative correlation between C_{org} and C_{mic}/C_{org} ratio and qM (r = -0.22 and r = -0.77, p < 0.001, respectively).

CONCLUSION

The variation of moisture content of the cover mixture during the experiment has been drastically affected by the climate conditions. In particular the low moisture content especially in the 4th sampling was a result of a prolonged period of heat and dryness. The differences between the plant species treatments that were observed in some samplings (3rd and 4th) can be attributed to the different vegetation growth rate, which correspondingly affects the moisture conditions of the substrate (Raich and Tufekcioglu 2000). Baldrian et al. (2010) showed that there is a considerable spatial variation of soil moisture, that can be partly attributed in the differences of the amount of rain that reaches the soil depending on the vegetation cover (Möttönen et al. 1999), but also in other processes such as diffusion of water, its uptake by roots or its displacement by mycelium hyphaes (Fricker et al. 2008) and its organic matter content (Fierer et al. 2009; Wardle 1992).

As the result analysis of the substrate properties suggests, the biology of the cover mixture changes over the 3.5 years period of the restoration, as it was also reported in a previous research (Alifragki et al. 2018).

The microbial biomass rate of change (ΔC_{mic}) is stable over time, with exception of the 4th sampling, when it decreased taking negative values in all treatments. In this particular sampling the moisture content of the mixture was extremely low compared to the rest. It is known that the microbial biomass shows seasonal variation (Bastida et al. 2006; Kamala and Sunada 2014; Masto et al. 2006), following the seasonal changes of soil moisture (Baldrian et al. 2010; Gömöryová et al. 2006; Stoyan et al. 2000). Furthermore, it has been found that the microbial biomass decreases significantly during dry periods and the moisture content correlates positively with C_{mic} (Niemeyer et al. 2012; Peigné et al.2009; Röver and Kaiser 1999) as has been found in this research as well.

The rate of CO_2-C release from basal respiration (ΔBR) in the rhizosphere of the three plant species increased until the 3rd sampling and decreased in the 4th, following the change of microbial biomass rate (ΔC_{mic}). The relationship between the above properties of the mixture was ascertained also by the existence of a significant positive correlation between the CO_2-C release and the C_{mic}, as similarly found by many other researchers (eg Chodak and Niklińska 2012; Dilly 2006, Garcia et al. 2005; Niemeyer et al. 2012). The rate of soil respiration as well as the microbial biomass depends on a large extent on the temperature and moisture conditions in the soil (eg Raich and Tufekcioglu 2000). The seasonal changes in the soil microclimate also play an important role in the seasonal CO_2 emission of an area (Nazaries et al. 2015). However, there was a negative correlation between respiration and moisture which can be probably attributed to the stress endured by microorganisms develop in these given conditions. Killham (1985) and Killham and Firestone (1984) showed that microorganisms consume more energy for their conservation than their growth, as stress is increased.

Microbial activity includes various metabolic processes that cannot be assessed by a single property only, but by a combination of them (Bastida et al. 2006; Nannipieri et al. 1990). It is known that C_{mic}/C_{org} ratio is suggested as indicator of disturbance or stress of the ecosystem (Wardle and Ghani 1995) and it depends strongly on the climatic conditions (Garcia et al. 2005; Insam 1990; Insam et al. 1989). Subsequently using those in the present study it has been ascertained that in conditions of water stress the values of C_{mic}/C_{org} ratio were significantly lower in all the plant treatments and also was found a positive correlation between C_{mic}/C_{org} and the moisture content of the substrate.

A useful tool in the understanding of microbial activity as also the metabolic quotient (qCO_2) as a means of "internal monitoring" of the substrate (Aceves et al. 1999; Joergensen and Brookes 2005). The study of qCO_2 clearly demonstrated that the values were larger when the moisture content was low. The dependence of qCO_2 on environmental factors like moisture including its seasonal changes is known (Šantr and SiraŠicraba 1991) and consolidated with the existence of a negative correlation between qCO_2 and moisture content. In this way, the variability of qCO_2 can show short term changes in substrate's conditions (Sinha et al. 2009) and can also be used as an indicator of microorganisms stress (Dilly and Munch 1998; Wardle and Ghani 1995). The higher the qCO_2 values, the lower the efficiency of the microbial activity and they live in stressful conditions (Sinha et al. 2009).

Alkaline phosphatase activity has shown seasonal variation as has been found by others (eg Alarcón-Gutiérrez et al. 2010; Baldrian et al. 2008), where it is claimed that it is affected by the seasonal changes in the moisture and temperature of the soil (Harrison 1983; Speir and Cowling 1991). In addition, a significant positive correlation was found between moisture content and phosphatase activity, as was also observed in similar studies (eg Alarcón-Gutiérrez et al. 2010; Gispert et al. 2013; Sardans and Peñuelas 2005).

The microbial activity in the rhizosphere of the three plant species that were established in the quarry area varied in time, depending on the growth rate and the characteristics of each species (Alifragki et al. 2018). To be more specific, it was found that during the first year of restoration in the rhizosphere of broom plants the microbial biomass (C_{mic}), the C_{mic}/C_{org} ratio, the CO_2-C release from basal respiration (BR) and the alkaline phosphatase activity (ALP) were statistically higher in relation to the rhizosphere of the other two plant species. Additionally, the metabolic quotient (qCO_2) in this treatment was lower but without showing significant differences.

The high microbial activity during the first year of broom establishment can be attributed to the fact that it is a legume shrub that develops a *Rhizobium* symbiosis and its roots secretions are rich in flavonoids, isoflavonoids and alkaloids (DiMichele et al. 2006; Pankhurst and Biggs 1980; Wink and Witte 1987). Moreover, it is known that broom is found in Eu-mediterranean and submediterranean vegetation zone, and has large adaptation ability in adverse environmental conditions, is installed in polluted soils (Muzzi and Fabbri 2007) and is used extensively in plantings for slope stabilization and corrosion protection due to the rapid development of its roots system.

Two years after the establishment of vegetation in the quarry space (3rd sampling), in pine's rhizosphere almost all biological properties of the cover mixture differed statistically significant compared to the other two plant species. Thus, C_{org}, microbial biomass (C_{mic}), C_{mic}/C_{org} ratio, CO_2-C release from basal respiration (BR), mineralization quotient (qM), alkaline phosphatase activity (ALP) were significantly higher and the metabolic quotient (qCO_2) was significantly lower.

In the 4th sampling, where the moisture content of the cover mixture was limited, the treatment of the cypress showed the largest values of its biological properties in its rhizosphere. Microbial biomass (C_{mic}), C_{mic}/C_{org} ratio, CO_2-C release from basal respiration (BR) and mineralization quotient (qM) were statistically significantly higher and the metabolic quotient (qCO_2) was significantly lower compared to the treatments of the other two plant species.

Pine as well as the cypress seemed to have slower microbial growth rate in its rhizosphere compared to the broom. Even though the development of pine root system is different from that of broom, the release of flavonoids in the rhizosphere from the secondary metabolic cycle can help in their growth and the increasing of microbial activity even in substrate with high concentrations of metals such as Mn, Fe, Cu (De Souza and de Giovani 2005; Fernandez et al. 2002; Mira et al. 2002; Özdemir 2005). However, it is known that *Pinus brutia* can be developed in most soil types and it is also tolerant to dryness and to low levels of nutrients in the soil (Quézel 2000). For these reasons *Pinus brutia* is widely used in reforestation. Furthermore, *Cupressus sempervirens* L. coexists with fungi of the *Glomeromycota* phyla forming AMF (arbuscular mycorrhizal fungi) mycorrhiza type (Ouahmane et al. 2006), thus contributing to the development of the rhizosphere's microbial community and gaining higher tolerance in dry conditions (Kyriazopoulos et al. 2014; Wu et al. 2013).

In the end, the development of the microbial activity in a substrate that consists mostly of pyrolusite by-products depends on the plant species that are established in the area to be restored in combination with abiotic factors such as moisture.

REFERENCES

Aceves, Martha, Carol Grace, Javier Ansorena, Luc Dendooven, and Philip Charles Brookes. 1999. "Soil microbial biomass and organic C in a gradient of zinc concentrations around a spoil tip mine." *Soil Biology and Biochemistry* 31:867–876.

Alarcón-Gutiérrez, Enrique, Carine Floch, Fabio Ziarelli, Christopher Augur, and Stéven Criquet. 2010. "Drying-rewetting cycles and γ-irridiation effects on enzyme activities of distrinct layers from *Quercus ilex* L. litter." *Soil Biology and Biochemistry* 42:283-290.

Alifragki, Maria G., Athina K. Pavlatou, and Michail Z. Orfanoudakis. 2018. "Phytoremediation affects microbial development on a limestone quarry." *International Journal of Phytoremediation* 20:957-963.

Anderson, Traute-Heidi, and Klaus Heinz Domsch. 1989. "Ratios of microbial biomass carbon to total organic carbon in arable soils." *Soil Biology and Biochemistry* 21:471-479.

Anderson, Traute-Heidi, and Klaus Heinz Domsch. 1990. "Application of eco-physiological quotients (qCO_2 and *q*D) to microbial biomasses from soils of different cropping histories." *Soil Biology and Biochemistry* 22:251–255.

Anderson, Traute-Heidi, and Klaus Heinz Domsch. 1993. "The metabolic quotient for CO_2 (qCO_2) as a specific activity parameter to assess the effect of environmental conditions, such as pH, on the microbial biomass of forest soils." *Soil Biology and Biochemistry* 25:393-395.

Baldrian, Petr, Josef Trögl, Jan Frouz, Jaroslav Šnajdr, Vendula Valášková, Věra Merhautová, Tomáš Cajthaml, and Jana Herinková. 2008. "Enzyme activities and microbial biomass in topsoil layer during spontaneous succession in spoil heaps after brown coal mining." *Soil Biology and Biochemistry* 40:2107-2115.

Baldrian, Petr, Věra Merhautová, Mirka Retránková, Tomáš Cajthaml, and Jarslav Šnajdr. 2010. "Distribution of microbial biomass and activity of extracellular enzymes in a hardwood forest soil reflect soil moisture content." *Applied Soil Ecology* 46:177-182.

Bastida, Felipe, José Luis Moreno, Teresa Hernandez, and Carlos Garcia. 2006. "Microbiological degradation index of soils in a semiarid climate." *Soil Biology and Biochemistry* 38:3463–3473.

Bauhus, Jürgen, David Parhé, and Leanadre Côté. 1998. "Effects of tree species, stand age and soil type on soil microbial biomass and its activity in a southern Boreal Forest." *Soil Biology and Biochemistry* 30:1077-1089.

Bertin, Cecile, Xiaohan Yang, and Leslie A. Weston. 2003. "The role of root exudates and allelochemicals in the rhizosphere." *Plant and Soil* 256:67-83.

Brookes, Philip Charles. 1995. "The use of microbial parameters in monitoring soil pollution by heavy metals." *Biology and Fertility of Soils* 19:269-279.

Chodak Marcin, and Maria Niklińska. 2012. "Development of microbial biomass and enzyme activities in mine soils." *Polish Journal of Environmental Studies* 21:569-577.

Criquet, Stéven. 2002. "Measurement and characterization of cellulase activity in sclerophyllous forest litter." *Journal of Microbiological Methods* 50:165–173.

de Souza, Rubens F.V., and Wagner F. De Giovani. 2005. "Synthesis, spectral and electrochemical properties of Al(III) and Zn(II) complexes with flavonoids." *Spectrochimica Acta Part A: Molecular and Biomolecular Spectroscopy* 61:1985-1990.

Dilly, Oliver. 2006. "Ratios of microbial biomass estimates to evaluate microbial physiology in soil." *Biology and Fertility of Soils* 42:241-246.

Dilly, Oliver, and Jean Charles Munch. 1998. "Ratios between estimates of microbial biomass content and microbial activity in soils." *Biology and Fertility of Soils* 27:374–379.

DiMichele, William, Neil Tabor, Das S., Chaney, and John W. Nelson. 2006. "From wetlands to wet spots: environmental tracking and the fate of Carboniferous elements in Early Permian tropical floras." In *Wetlands through time* edited by Stephen F. Greb, and William A.

DiMichele, 223–248. USA: Geological Society of America, Special Paper 399.

Fernandes, Silvana, Aparecida Pavan, Wagner Bettiol, and Carlos Clementi Cerri. 2005. "Effect of sewage sludge on microbial biomass, basal respiration, metabolic quotients and soil enzymatic activity." *Applied Soil Ecology* 30:65-77.

Fernandez, Tereza M., Lurdes M. Mira, Helena M. Florêncio, and Keith R. Jennings. 2002. "Iron and copper chelation by flavonoids: an electrospray mass spectrometry study." *Journal of Inorganic Biochemistry* 92:105-111.

Fierer, Noah, Michael S. Strickland, Daniel Liptzin, Mark A. Bradford, and Cory C. Cleveland. 2009. "Global patterns in belowground communities." *Ecology Letters* 12:1238-1249.

Fricker, Mark D., Dan Bebber, and Lynne Boddy. 2008. "Mycelial networks: structure and dynamics." In *Ecology of Saprotrophic Basidiomycetes,* edited by Lynne Boddy, Juliet C Frankland, and Pieter van West, 3–18. New York: Academic Press.

Garcia, Carlos Andrés, Antonio Roldán, and Teresa Hernandez. 2005. "Ability of different plant species to promote microbiological processes in semiarid soil." *Geoderma* 124:193-202.

George, Timothy S., Peter J. Gregory, Martin Wood, Derek Read, and Roland J. Buresh. 2002. "Phosphatase activity and organic acids in the rhizosphere of potential agroforestry species and maize." *Soil Biology and Biochemistry* 34:1487-1494.

Giller, Ken E., Ernst Witter, and Steve P. McGrath. 1998. "Toxicity of heavy metals to microorganisms and microbial processes in agricultural soils: a review." *Soil Biology and Biochemistry* 30:389-1414.

Gispert, Marina, Mohamed Emran, Giovanni Pardini, Serena Doni, and Brunello Ceccanti. 2013. "The impact of land management and abandonment on soil enzymatic activity, glomalin content and aggregate stability." *Geoderma* 202-203:51-61.

Gömöryová, Erika, Juraj Gregor, Viliam Pichler, and Dušan Gömöry. 2006. "Spatial patterns of soil microbial characteristics and soil moisture in a natural beech forest." *Biologia* 61:S329–S333.

Graham, Michael H., and Richard J. Haynes. 2004. "Organic matter status and the size, activity and metabolic diversity of the soil microflora as indicators of the success of rehabilitation of sand dunes." *Biology and Fertility of Soils* 39:429-437.

Haripal, Kamala, and Sunada Sahoo. 2014. "Microbial biomass carbon, nitrogen and phosphorus dynamics along a chronosequence of abandoned tropical agroecosystems." *International Journal of Current Microbiology Applied Sciences* 3:956-970.

Harris, Jim Arthur. 2003. "Measurements of the soil microbial community for estimating the success of restoration." *European Journal of Soil Science* 54:801-808.

Harrison, Anthony F. 1983. "Relationship between intensity of phosphatase activity and physico-chemical properties in woodland soils." *Soil Biology and Biochemistry* 15:93-99.

Insam, Heribert. 1990. "Are the soil microbial biomass and basal respiration governed by the climatic regime?" *Soil Biology and Biochemistry* 22:525-532.

Insam, Heribert, Dennis Parkinson, and Klaus Heinz Domsch. 1989. "Influence of macroclimate on soil microbial biomass." *Soil Biology and Biochemistry* 21:211-221.

Insam, Heribert, and Klaus Heinz Domsch. 1988. "Relationship between soil organic carbon and microbial biomass chronosequences of reclamation sites." *Microbial Ecology* 15:177-188.

Joergensen, Rainer Georg, and Philip C Brookes. 2005. "Quantification of soil microbial biomass by fumigation-extraction." In *Monitoring and Assessing Soil Bioremediation. Soil Biology, vol 5*, edited by Rosa Margesin, and Franz Schinner, 281-295. Berlin- Heidelberg: Springer.

Killham, Kenneth S. 1985. "A physiological determination of the impact of environmental stress on the activity of microbial biomass." *Environmental Pollution* 38:283-294.

Killham, Kenneth, and Mary K. Firestone. 1984. "Salt stress control of intracellular solutes in streptomycetes indigenous to saline soils." *Applied and Environmental Microbiology* 47: 301-306.

Kyriazopoulos, Apostolos P., Michail Orfanoudakis, Eleni M. Abraham, Zoi M.M. Parissi, and Nikoleta Serafidou. 2014. "Effects of arbuscular mycorrhiza fungi on growth characteristics of *Dactylis glomerata* L. under drought stress conditions." *Notulae Botanicae Horti Agrobotanici Cluj-Napoca* 42:132-137.

Land Use Consultants. 1992. *Amenity reclamation of mineral working; main report*. London: Crown.

Legwaila, Israel A., Eckart Lange, and John Cripps. 2015. "Quarry reclamation in England: A review of techniques." *Journal of the American Society of Mining and Reclamation* 4:55-79.

Luna, Lourdes, Roberta Pastorelli, Felipe Bastida, Teresa Hernández, Carlos García, Isabel Miralles, and Albert Solé-Benet. 2016. "The combination of quarry restoration strategies in semiarid climate induces different responses in biochemical and microbiological soil properties." *Applied Soil Ecology* 107:33-47.

Masto, Ebhin R., Pramod K. Chhonkar, Dhyan Singh, and Ashok K. Patra. 2006. "Changes in soil biological and biochemical characteristics in a long-term field trial on a sub-tropical Inceptisol." *Soil Biology and Biochemistry* 38:1577-1582.

Mira, Lurdes, Tereza M. Fernandez, Marta Santos, Rui Rocha, Helena M. Florêncio, and Keith Jennings. 2002. "Interactions of Flavonoids with Iron and Copper Ions: A Mechanism for their Antioxidant Activity." *Free Radical Research* 36:1199-1208.

Moscatelli, Maria Cristina, Alessandra Lagomarsino, Sara Marinari, Paolo De Angelis, and Stefano Grego, S. 2005. "Soil microbial indices as bioindicators of environmental changes in a poplar plantation." *Ecological Indicators* 5:171-179.

Möttönen, Marjoriitta, Erkki Jarvinen, Timo J. Hokkanen, Timo Kuuluvainen, and Rauni Ohtonen. 1999. "Spatial distribution of soil ergosterol in the organic layer of a mature Scots pine (*Pinus sylvestris* L.) forest." *Soil Biology and Biochemistry* 31:503-516.

Muzzi, Enrico, and Fabbri, T. 2007. "Revegetation of mineral clay soils: shrub and tree species compared." *Land Degradation and Development* 18:441–451.

Nannipieri, Paolo, Stefano Grego, and Brunello Ceccanti. 1990. "Ecological significance of the biological activity in soils." In *Soil Biochemistry volume 6*, edited by Jean-Marc Bollag, and G. Stotzky, 293–355. New York, NY: Marcel Dekker.

Nazaries, Loïc, William Tottey, Lucinda Robinson, Amit Khachane, Waleed Abu Al-Soud, SørenSørensen, and Brajesh Singh. 2015. "Shifts in the microbial community structure explain the response of soil respiration to land-use change but not to climate warming." *Soil Biology and Biochemistry* 89:123-134.

Nelson Darrell W., and Lee Sommers. 1982. "Total carbon, organic carbon and organic matter." In *Methods of Soil Analysis: Part 2 Chemical and Microbiological Properties,* edited by Albert L. Page, 539-579. Madison: American Society of Agronomy.

Niemeyer, Júlia Carina, Giovana Bortoti Lolata, Gabriel Martins de Carvalho, Eduardo Mendes Da Silva, José Paulo Sousa, and Marco Antonio Nogueira. 2012. "Microbial indicators of soil health as tools for ecological risk assessment of a metal contaminated site in Brazil." *Applied Soil Ecology* 59:96-105.

Ouahmane, Lahcen, Mohamed Hafidi, Christian Plenchette, Marija Kisa, Ali Bournezzough, Jean Thioulouse, and Robin Duponnois. 2006. "*Lavandula* species as accompanying plants in *Cupressus* replanting strategies: Effect on plant growth, mycorrhizal soil infectivity and soil microbial catabolic diversity." *Applied Soil Ecology* 34:190-199.

Özdemir, Zeynep. 2005. "*Pinus brutia* as a biogeochemical medium to detect iron and zinc in soil analysis, chromite deposits of the area Mersin, Turkey." *Geochemistry* 65:79-88.

Pandey, Anita, and Lok Man S. Palni 2007. "The rhizosphere effect in trees of the Indian Central Himalaya with special reference to altitude." *Applied Ecology and Environmental Research* 5:93-102.

Pankhurst, Caroline E., and David R. Biggs. 1980. "Sensitivity of *Rhizobium* to selected isoflavonoids." *Canadian Journal of Microbiology* 26:542-545.

Pascual, Jose Antonio, Carlos Garcia, Teresa Hernandez, José Moreno, and Margarita Ros. 2000. "Soil microbial activity as a biomarker of degradation and remediation processes." *Soil Biology and Biochemistry* 32:1877-1883.

Peigné, Joséphine, Jean – François Vian, Mario Cannavacciuolo, B. Bottollier, and RémiChaussod. 2009. "Soil sampling based on field spatial variability of soil microbial indicators." *European Journal of Soil Biology* 45:488-495.

Pinzari Flavia, Alessandra Trinchera, Anna Benedetti, and Paolo Sequi. 1999. "Use of biochemical indices in the Mediterranean environment: comparison among soils under different forest vegetation." *Journal of Microbiological Methods* 36:21–28.

Quézel, Pierre. 2000. "Taxonomy and biogeography of Mediterranean pines (*Pinus halepensis* and *P. brutia*)." In *Ecology, biogeography and management of Pinus halepensis and P. brutia forest ecosystems in the Mediterranean basin,* edited by Gidi Néeman, and Louis Trabaud, 1-12. Netherlands: Backhuys, Leiden.

Raich, James W., and Aydin Tufekcioglu. 2000. "Vegetation and soil respiration: correlations and controls." *Biogeochemistry* 48:71–90.

Roldán, Antonio, Fuensanta Garcia-Orenes, and Antonio Lax. (1994). "An incubation experiment to determine factors involving aggregation changes in an arid soil receiving urban refuse." *Soil Biology and Biochemistry* 26:1699-707.

Röver, Manuela, and Ernst-August Kaiser. 1999. "Spatial heterogeneity within the plough layer: low and moderate variability of soil properties." *Soil Biology and Biochemistry* 31:175–187.

Rowell David L. 1994. *Soil Science: Methods and Applications*. UK: Longman Group.

Šantrůčková, Hana, and Milan Straškraba. 1991. "On the relationship between specific respiration activity and microbial biomass in soils." *Soil Biology and Biochemistry* 23:525-532.

Sardans, Jordi, and Josep Peñuelas. 2005. "Drought decreases soil enzyme activity in a Mediterranean *Quercus ilex* L. forest." *Soil Biology and Biochemistry* 37:455-461.

Saviozzi, Alessandro, Renato Levi-Minzi, Roberto Cardelli, and Riccardo Riffaldi. 2001. "A comparison of soil quality in adjacent cultivated, forest and native grassland soils." *Plant and Soil* 233:251-259.

Sinha, Shipra, Reginald Masto, Lal Ram, Vetrivel A. Selvi, Nishant Kumar Srivastava, Ramesh C. Tripathi, and Joshy George. 2009. "Rhizosphere soil microbial index of tree species in an coal mining ecosystem." *Soil Biology and Biochemistry* 41:1824-1832.

Sparling, Graham P., Phil Hart, Jessica A. August, and David M. Leslie. 1994. "A comparison of soil and microbial carbon, nitrogen, and phosphorus contents, and macro-aggregate stability of a soil under native forest and after clearance for pastures and plantation forest." *Biology and Fertility of Soils* 17:91-100.

Speir, Thomas W., and John C. Cowling. 1991. "Phosphatase activities of pasture plants and soils: relationship with plant productivity and P fertility indices." *Biology and Fertility of Soils* 12:189-194.

Stoyan, Helmut, Helvecio De-Polli, Sven Böhm, Philip G. Robertson, and Eldor A. Paul. 2000. "Spatial heterogeneity of soil respiration and related properties at the plant scale." *Plant and Soil* 222:203–214.

Stroo, Hans F., and Everett M. Jencks 1982. "Enzyme activity and respiration in minesoils." *Soil Science Society of America Journal* 46:548-553.

Tabatabai, M. Ali. 1994. "Soil enzymes." In *Methods of Soil Analysis" Part 2 Microbiological and Biochemical Properties,* edited by Peter J. Bottomley, J. Scott. Angle, and R. W. Weaver, 775-859. Madison, WI: SSSA Book Series, Soil Science Society of America.

Tang, Xuli, Shuguang Liu, Guoyi Zhou, Deqiang Zhang, and Cunyu Zhou. 2006. "Soil-atmospheric exchange of CO_2, CH_4, and N_2O in three subtropical forest ecosystems in southern China." *Global Change Biology* 12:546–560.

Tordoff, George M., Alan John Martin Baker, and Arthur J. Willis. 2000. "Current approaches to the revegetation and reclamation of metalliferous mine wastes." *Chemosphere* 41:219-228.

Wardle, David A. 1992. "A comparative assessment of factors which influence microbial biomass carbon and nitrogen levels in soil." *Biological Reviews* 67:321–358.

Wardle, David A., and Anwar Ghani. 1995. "A critique of the microbial metabolic quotient (qCO_2) as a bioindicator of disturbance and ecosystem development." *Soil Biology and Biochemistry* 27:1601-1610.

Wink, Michael, and Ludger Witte. 1987. "Alkaloids in Stem Roots of *Nicotianatabacum* and *Spartium junceum* Transformed by *Agrobacterium rhizogenes.*" *Zeitschriftfür Naturforschung C* 42:69-72.

Xu, Qiang-Sheng, Anoop Kumar Srivastava, and Ying-Ning Zou. 2013. "AMF-induced tolerance to drought stress in citrus: A review." *Scientia Horticulturae* 164:77-87.

In: The Future of Phytoremediation
Editor: James C. Flores
ISBN: 978-1-53619-625-2

Chapter 6

PHYTOREMEDIATION AND PURPLE NON SULFUR BACTERIA

Muazzez Gürgan[1] and Sevinç Adiloğlu[2,*]
[1]Department of Biology, Tekirdağ Namık Kemal University, Tekirdağ, Turkey
[2]Department of Soil Science and Plant Nutrition, Tekirdağ Namık Kemal University, Tekirdağ, Turkey

ABSTRACT

Microorganisms are vital for agriculture as they possess biochemical reactions required for the quality and productivity of agricultural soils. Bacteria are the most diverse and intense group of microorganims and essential members of soil microorganisms as they produce phytohormones, secondary metabolites and protect plants against diseases. Bacterial diversity and amount is crucial for cycling of elements and nutrients in soil, therefore are one of the factors determining the fertility of soil. Purple non sulfur bacteria are a non taxonomic group of bacteria which are Gram negative bacteria having many growth modes. They enrich soil by fixing carbon and atmospheric nitrogen and making

* Corresponding Author's E-mail: sadiloglu@hotmail.com.

them available for plant uptake. This way these bacteria can be employed as biofertilizers in agriculture to enhance growth of different ground and surface plant parts. This review summarizes the properties of purple non sulfur bacteria and their use in plant growth enhancement and bioremediation.

Keywords: purple non sulfur bacteria, plant growth promoter, phytohormones, bioremediation

INTRODUCTION

Soil is a live, dinamic and three-dimensional environment. Microorganisms constitute the greatest portion of the living part of the soil. Bacterial population is the most crowded among the microorganisms in the soil (Tok 1998). Their amount in soil is a few billion/1g soil in general. However, the total weight and volume occupied are so small as the size of them usually are not larger than a few microns. For this reason, they make up almost half of the total microorganism weight in soil. Bacteria are the dominating microorganisms in the well aerated soils. They can be classified systematically, taxonomically as well as physiologically and metabolically. Criteria such as energy sources, carbohydrates, nitrogen utilization are taken into account in physiological and metabolic classification. Bacteria especially play role physiologically in nitrogen, carbon and iron cycle (Delgado-Baquerizo et al. 2018).

Under harsh soil environmental conditions such as freezing-thawing, drying-soaking and high-low temperatures, bacteria transform into endospore form and can turn back into vegetative form when the environmental conditions reach optimum (Kennedy et al., 1994).

The most intense soils in terms of bacterial population diversity are that used to grow legumes. Bacterial amount and mass are also high in the soil under grassland (Zhou et al. 2017). Bacterial cell number is in average hundred times more than that of fungi, but they have less mass than fungi in total. However, bacterial populations produce more biochemical energy.

The bacterial diversity and amount depend on the soil characteristics and applications on soil, and they increase towards the warm climates. The organic content and ecological conditions are the determinant factors on bacterial cell number in the soil. Cultivated soils contain more bacteria than virgin soils in the same region (Siles and Margesin 2016).

The fertilizers applied on the soil being organic or inorganic origin affect the bacterial diversity and number. Also, the variety of the fertilizer plays role on increasing or decreasing bacterial amount in soil. For example, addition of a fertilizer with ammonium causes a decrease in the bacterial population in soil as the hydrogen ions generated during the conversion of ammonium ions into nitrate ions decrease the pH and asidifies the soil environment. In fact, most bacteria grow in neutral pH. The bacterial number decreases rapidly in acidic or alkali soils. Particularly, liming of acidic soils increases the bacterial population in soil (Adiloğlu et al., 2018; Adiloğlu and Adiloğlu, 2005). The other factors affecting bacterial number and species are soil processing, humidity and aeration of soil, soil texture and depth, temperature, precipitation, vegatation, plant roots and seasonal changes.

As a result of conventional agriculture, soil micro-organisms are decreasing and soil fertility is decreasing. However, the soils that form the basis of human nutrition are in serious danger with pollution. Soils are polluted in many ways, especially by wrong agricultural activities. In recent years, phytoremediation, the most economical and applicable technology, has come in pollution removal. With this study, the promising method in the removal of soil pollution, phytoremediation and Purple non sulfur bacteria, which has an important role in soil fertility, will be evaluated together (Adiloğlu 2016; Adiloğlu 2017; Adiloğlu and Göker 2021; Adiloğlu 2021).

USE OF BACTERIA AS BIOFERTILIZER

Chemical fertilizers have been applied to increase soil fertility in modern agriculture practices. They contain nitrogen and phosphorus as primary components. The soil chemical, physical and biological properties can change upon usage of chemical fertilizers which decrease soil microbial population, therefore crop productivity decreases, as soil microoganisms maintain and improve soil fertility by the biological reactions they carry out (Hu et al. 2011). In order to maintain the sustainability in soil productivity and quality, one of the most important strategies is increasing beneficial bacterial populations using bacteria as biofertilizer (Karaman 2012). Biofertilizers are preparations that include live and latent bacteria which can fix nitrogen, phosphate and some other plant micronutrients and make them available for plant uptake, and also promote plant growth by production of secondary metabolites and protect plants against some diseaseses. The effective microorganisms in the biofertilizers are actinomycetes, lactic acid bacteria, photosynthetic bacteria and some yeasts (Daly and Stewart 1999). The benefits of inoculating effective microorganisms to the soil come from their production of antioxidants, plant hormones (auxin, giberellin, absisic acid and cytokinin) which induce root and shoot growth and solubilization of nutrients to make them available for the plants (Karadeniz & Topcuoglu, 2006; van der Heijden et al., 2008; Verbon & Liberman, 2016). The biofertilizers can be applied on seeds, soils, plant wastes, and composts in order to increase the rates of microbial activities which enhance the effectiveness of plant nutrient elements required for plant growth (Sheraz Mahdi et al. 2010).

These bacteria that enhance plant growth are classified in two groups. The first are the bacteria which can biologically fix nitrogen, solubilize inorganic phosphate and iron, produce the plant growth promoters such as phytohormones, vitamins and enzymes, and increase the resistance of plants against saltiness, drought, pesticides and metal toxicity (Bowen and Rovira 1999; Karaman 2012). The other group contain the bacteria which can decrease or inhibit phytopathogenic microorganisms (Canbolat et al. 2006; Karaman 2012). Applications of plant growth promoters enhance root growth in a very short time and and provides protection. Especially during the early seedling phase when the sensitivity for ambient conditions is high, the plants can easily survive stress factors and can develop in a healthy way. The biofertilizers have recently been applied on sugar beet, sugar cane, rice, maize and wheat (Çakmakçı et al. 2006; Karaman 2012).

The substances such as the mentioned hormones that induce plant growth are also called plant growth promoters. These substances can be produced by plants and microorganisms. Those microorganisms living in the soil around the roots of the plants, and directly or indirectly affect plant growth are called plant growth promoter (rhizo) bacteria (Verma et al. 2010). Indole 3-acetic acid (IAA) and 5-aminolevulinic acid (ALA) are typical substances generated by plant growth promoters that induce plant growth. IAA is a phytohormone of the auxin class that affects cell enlargement, cell division, growth rate, root initiation, phototropism, geotropism and apical dominance (Frankenberger and Arshad, 1995). The bacteria interacting with plants can produce high amount of IAA and can change the local phytohormone balance of the plant (Glick et al., 1999) ALA is a universal intermediate in chlorophyll, heme and Vitamin B12 synthesis (Hotta et al. 1997, Kang et al. 2012). The effects of ALA have been investigated in many researches and it was revealed that ALA is not harmful to crops, humans and animals, and therefore can be used as an effective herbicide. Besides chlorophyll production, photosynthesis, and plant growth, it also enhances plant stress tolerance (Hotta et al. 1997, Yang et al. 2014).

PURPLE NON SULFUR BACTERIA

Purple non sulfur bacteria (PNSB) are a non taxonomic group formed by the species of alpha and beta-proteobacteria (Imhoff 2006). They are photosynthetic gram-negative bacteria which perform anoxygenic photosythesis and convert light energy to chemical energy. They possess bacteriochlorophylls and carotenoids as the photosynthetic pigments. They differ from purple sulfur bacteria by utilization of sulfide: purple sulfur bacteria can utilize sulfide and store sulfur in intercellular globules, however purple non sulfur bacteria do not store elemental sulfur (Madigan and Jung 2009). Photosynthesis by PNSB occur under oxygen free conditions unlike cyanobacteria and aerobic anoxygenic phototrophs, because pigment synthesis is repressed by oxygen in these bacteria.

Morphologies of PNSB can be of various such as coccus, rod, sprilla and vibrio, belonging to twenty genera (Madigan and Jung 2009). *Rhodobacter capsulatus*, *Rhodobacter spaheroides*, *Rhodosprillum rubrum*, *Rhodosprillum centenum*, *Rhodopseudomonas palustris*, *Rhodobium marinum*, *Rubrivivax gelatinosa*, *Rhodobacter blasticus*, *Rhodobacter sulphidophilus* can be given as well-known examples of PNSB.

PNSB have various light dependent and independent growth modes such as aerobic respiration, anaerobic respiration, fermentation, photoautotrophy and photoheterotrophy (Koku et al. 2002). These accomplished bacteria have been an interest of research on carbon, nitrogen and energy metabolism (Androga et al. 2012). They are easy to culture in the laboratory as they need anoxic mineral medium with addition of either sulfide and bicarbonate for photoautotrophic growth or organic compound for photoheterotrophic growth, and the latter is the preferred growth mode by PNSB (Sojka 1978, Madigan & Jung 2009).

Purple non sulfur bacteria have been known and studied for their nitrogen fixation and hydrogen production capabilities in recent decades (Weaver et al. 1975). The main reason they can produce hydrogen is indeed the reason for their nitrogen fixing ability: the nitrogenase enzyme. By the help of this enzyme, bacteria can fix the free nitrogen in the air to supply the plants with their nitrogen needs. This property makes purple non sulfur bacteria enrich soil in nitrogen content.

PNSB as Plant Growth Promoter

The effects of PNSB as plant growth promoter microorganisms were tested on several plants, one of which was tomato. When PNSB were applied with a mineral fertilizer, the number of fruits of tomato increased (Ncube et al. 2016). Similar results of plant growth promoters on tomato were obtained as fruit number and vitamin C concentration of tomato were increased when they were added to an organic fertilizer Moreover, soil and leaf nitrogen contents increased with the application of these microorganisms. However, when they were applied alone, they were effective on increasing fruiting but not protecting fruits against fungi infection. Yet, the use of fungiside should be taken into account (Xu et al. 2001). In another study with tomato, the seeds were germinated with the addition of PNSB isolated from a river sediment in South Korea. First, ALA and IAA production of different *Rhodopseudomonas* strains were measured, then the highest producer strain was applied on seeds for germination. The whole lengths of seedlings, root and shoot lengths and dry mass were higher when PNSB were applied. The results suggested that PNSB produced phytohormones like the rhizobacteria but they increased the germination rate more than other typical plant growth rhizobacteria (Koh and Song 2007). The effects of PNSB on plant growth of tomato were tested in greenhouse conditions in another study. Again, the agro-morphological traits such as dry weight, shoot and root length, fresh weight of tomato fruits and lycopene content were higher with addition of PNSB. Moreover, it was determined that the native bacterial community in soil was not affected and disturbed by the added PNSB (Lee et al. 2008).

Rice was another plant on which PNSB were shown to have plant growth promoter effects. *Rhodobacter capsulatus* DSM155 enhanced dry weight, shoot height and N content of roots of seedlings of four rice varieties in hydroponic cultures. It was suggested that atmospheric nitrogen fixation ability of *R. capsulatus* made the rice seedlings to grow (Elbadry and Elbanna 1999). The same stain was also tested on rice in pot experiments with or without two different levels of N fertilizer (Elbadry et al. 1999). Inoculation of only bacteria increased plant height, panicle length and weight, grain and straw yield of rice significanlty. Moreover, the effect of inoculation of bacteria together with half amount of N fertilizer was compared to full amount of N fertilizer. The plant height, grain yield and harvest index (grain yield/above ground total plant biomass) were not significantly different. This showed that inoculation of a PNSB can decrease the need for chemical fertilizer usage. After these studies, field experiments were carried out with the application of *R.*

capsulatus DSM155 to rice pads. The inoculation of these bacteria significantly increased shoot height, grain yield, N contents of grains and straws in all the treatments (Gamal-Eldin and Elbanna 2011). In a more recent study, three *Rhodopseudomonas palustris* strains were shown to increase rice yields in both organic and saline flooded paddy fields (Kantachote et al. 2016). All PNSB produced maximum grain yields in saline fields however only one strain had a significant advantage in organic paddy field. It was suggested that PNSB reduced toxicity of sodium ions to rice root by entraping the available Na^+ in the saline soil and could help resisting the salty environment. Moreover, ALA produced well by PNSB reduces salt stress for plants and inhibited growth of methanogens. The methane emission of the PNSB isolated from saline paddy fields was reduced up to 88%. As paddy fields are important sources for methane production, which has a huge global warming potential, use of PNSB in paddy fields, besides enhancing plant growth, can reduce enviromental hazard of the paddy fields (Nunkaew et al. 2015, Kantachote et al. 2016).

Among the PNSB, *Rhodopseudomonas palustris* is typical one which can promote plant growth by producing ALA and riboflavin. This bacterium also produces siderophores, extracellular polymeric substances (EPS) and bacterial acyl-homoserine lactone (AHL) which can be interest of several researches (Sasaki et al. 2005). The photosynthetic *R. palustris* also fixes carbon and secretes up to 40% of the fixed carbon into the rhisozosphere, therefore has an additional effect on plant growth (Bais et al. 2006). Moreover, PNSB, unlike other plant growth promoters, can harvest light energy and use carbon dioxide as carbon source, thanks to the photoautotrophic growth mode (Simmons et al. 2011). This property makes PNSB grow on leaf surface. The application of PNSB to soil should be repeated many times to acquire a sufficient population size (Ferron and Deguine 2005). PNSB were thought to act in a similar mode and induce growth, systemic resistance and contribute to the health of the plant when applied to the phyllosphere, i.e., around the leaves, instead of rhizosphere (Berg 2009). Foliar spray of PNSB was shown to be an easy and efficient way of PNSB use in agriculture. The spray of *R. palustris* on the leaves of Chinese cabbage, precautiously not to contaminate the soil, slightly changed the soil community and accelerated carbon turnover and soil nutrient cycling. Moreover, this approach supressed the pathogens and increased the number of plant growth promoters, bacteria that fix nitrogen, and also the group having antifungal acitivity in the soil (Xu et al. 2016). *R. palustris* in the mentioned study was also grown on stevia residue wastewater which was rich in organic components. This wastewater usage for bacterial growth enhanced the production of ALA of *R. palustris*, photosynthetic rate and plant biomass. This study not only showed the plant growth promoter effects of PNSB, but also suggested an economic aspect of PNSB usage in agriculture as it was combined with wastewater treatment to decrease environmental harm of the wastewater rich in organics. Spray of *R. palustris* on the leaves of plant was also tested againt tobacco mosaic virus (Su et al. 2017). Besides the production of ALA and IAA of *R. palustris* increases the seed germination and dry mass of the plant, increase in the antioxidant enzyme activity and resistance of tobacco against tobacco mosaic virus was shown in this study. The production of IAA by *R. palustris* was shown in a study where *R. palustris* isolated from rice paddy fields in Taiwan was tested on Chinese cabbage *Brassica rapa chinensis* (Wong et al. 2014). *R. palustris* did not exhibit plant growth promoter effect on Chinese cabbage by itself, but when the chemical fertilizer amount was decreased by 50% and applied together with the bacteria, the effect was the same as 100% chemical fetilizer usage on dry and fresh weights of shoots. One important finding of this study was that PNSB were effective on old seeds with poor quality. When old seeds were used for germination with the addition of *R. palustris*, growth rates were greater with old seeds than new seeds. Therefore, old seeds can be evaluated with addition of PNSB suggesting a contribution agricultural economy.

PNSB in Phytoremediation

Human activities towards industrialization and urbanisation brought about many outputs such as wastewaters, excessive use of fuels, pesticides and fertilizers, which unavoidably release hazardous organic and inorganics (Ramanjaneyulu and Giri 2006). They eventually pollute water sources, atmosphere and the soils. Organic contaminants may be converted to carbon oxides by microbial activities in the environment, while most inorganics, especially heavy metals cannot be degraded chemically or microbiologically (Kirpichtchikova et al. 2006). Soils are the major sink for by heavy metals whose accumulations threaten the human and ecosystem through direct contact, food chain, reduction in agricultural land use and eventually food quality (McLaughlin et al., 2000). The hevay metals frequently contaminate the soils are arsenic, cadmium, chromium, copper, lead, mercury, nickel and zinc (Evanko and Dzombak 1997). Some research on use of PNSB to vitiate and clean up the heavy metals that causes problems in agricultural activities are given in this review.

For bioremediaiton of arsenic contaminated lands, As-resistant bacteria were suggested to be employed as As-resistant bacteria can methylate the toxic inorganic arsenic into less toxic methylated form thanks to the activity of arsM gene encoding (As (III)) S-adenosylmethyltransferase enzyme. The arsM gene of Rhodopseudomonas palustris has a high rate of methylation. The transformation of this gene from R. palustris to a Pseudomonas putida KT2440, which can succesfully reduce arsenate to arsenite and export arsenite to external environment but cannot methylate arsenic, enabled the latter strain to volatilize more arsenic from contaminated soil (Chen et al. 2014). The ability of oxidation/reduction of arsenic in soil by PNSB can therefore decrease bioavailability of as to plants. As-resistant PNSB isolated from a fish pond were checked for auxin production and plant growth promotion on Vigna mungo plant seeds in pot experiment (Batool et al., 2017). When the plants were solely exposed to As, root and shoot lengths decreased. However, when they were exposed to as together with PNSB treatment, both shoot and root lengths increased compated to the control. PNSB enhanced plant groeth both in absence and presence of arsenic.

The investigation of arsenic resistant mechanism in 4 PNSB strains (3 strains of *Rhodopseudomonas palustris* and one strain of *Rubrivivax benzoatilyticus*) revelaed that the most arsenic resistant bacterium was *R.*

palustris C1 strain which had all the As-marker genes: *arsC*, *arsM*, *aioA* and *acr3* (arsenate reductase, arsenite S-adenosylmethyltransferase, arsenite oxidase, arsenite transporter gene, respectively) Therefore this bacterium was suggested to have a high potential for bioremediation of arsenic contaminated area. (Nookongbut et al., 2017).

The *arsM* gene is present in some bacterial strain, however plants lack this gene and related enzyme. The first study in planta methylation and volatilizaiton of arsenic in higher plants was carried out by the expression of *arsM* gene of *R. palustris* in *Oryza sativa* (Japonica rice) which enabled the plant to methylate inorganic arsenicto a variety of organic species. Transgenic rice accumulated monomethylarsenate and dimethylarsenate in the roots and shoot. Moreover, seeds of transgenic rice had significantly lower As(III) and AS(V) in both seeds and husks compared to nontransgenic rice seeds (Meng et al., 2011).

Some micronutritious elements such as Cu and Zn are commonly present in the soil. However, their concentration can be increased due to pollution and can reach a toxic level for organisms (Fekiacova, Cornu, and Pichat 2015). The soil around the Hitachi mine, Japan, contains high number of heavy metals such as Zn, Cu, Pn and Cd which makes it difficult for the plants to grow. However, a perennial helophyte Cicuta virosa L. plant can grow in a small pond at the mine. This plant can accumulate high concentrations of Zn in the roots in an amount more than 20 times enought to reach toxic levels of crops. The heavy metal concentrations in the pond water were below detection limits but very high in the pond sediment. Nagata et al. (2015) isolated two endophytes, *Pseudomonas putida* and *Rhodopseudomonas* sp. from the pond. They grew the seeds of Cicuta virosa L. in sterilized pond sediment and sterillized water inoculated with 107/mL of either one of these endophytes and compared their number of leaves, seedling height, freah and dry mass of above ground parts, fresh and dry mass of roots. All these traits were much higher than those of control condition that was inoculated wth autoclaved bacterial suspension. The seedlings grown in the absence of endophyte showed signals of hevay metal toxicity. The bacteria incubated solubilized Zn from sediment by production of Zn chelating compounds and make the plant uptake Zn. These bacteria also detoxify heavy metal in root tissue and enhance growth of C. virosa L seedlings in the presence of heavy metals, possibly by inducing indole-3-acetic acid in the plant.

PNSB were also isolated from various paddy fields some of which were contaminated with Cd and Zn, as they were sound mine areas. Sakpirom et al. (2017) found a strain that produced high amount of ALA and IAA (4.11 and 3.62 mg/L, respectively), which turned out to be Rhodobacter palustris. This strain also removed 85 and 55% of Cd and Zn. They found another strain producing 0.94 mg/L ALA and 0.65 mg/L IAA which could remove 72 and 74% of Cd and Zn of the samples contaminated with the same amount of Zn and Cd as the paddy fields. They detected that this latter strain was Rubrivivax gelatinosus, which is a PNSB. *R. gelatinosus* was previously found to be more resistant to heavy metal contamination of Pb, Hg and Cr than two other PNSB: *R. rubrum* and *R. sphaeroides* (Kis et al., 2015).

In another study, when cucumber seedlings were cotreated with different concentrations of Cd and *R. palustris* at the same time, *R. palustris* significantly lowered the toxic effects of Cd on cucumber while Cd reduced the shoot and root length, total chlorophyll content and root activity to uptake water and nutrients (Ge, Liu, and Zhang 2017). In addition, the bacteria also induced soluble sufar and protein content in cucumber seedlings both with and without Cd treatment. Moreover, it also caused production of antioxidant enzymes SOD and POD in both control and Cd treatment. This helps plants against stress condition and to adapt changing environment. This study also revealed the plant growth promoter effect of *R. palustris* on cucumber which can help cucumber in Cd containated soil.

Wastes of various industries including fuel and petrochemical industry and also mining sites can contain cobalt and nickel which may contaminate soil and water sources nearby and are toxic to life. (Harasim and Filipek 2015; James and Gibb 2006). Plants can be used for remediation of Co, however this process is not satisfactory as plants grow slowly with low biomass and their geographical distribution is restricted. PNSB possess Co-Ni permease enzyme which was cloned from *R. palustris* and introduced into tobacco plants. The transgenic tobacco accumulated significantly more Co in both shoots and roots compared to the control nontransgenic tobacco. This genetic modification also stimulated the accumulation of nickel, however cobalt accumulation was more preferred than nickle accumulation. It was suggested that in sites contamintated with Co and Ni, transgenic non-food, non-feed self pollinationg tobacco plants carrying Ni-Co permease of PNSB can be used for remediaiton of these metals (Nair et al. 2012).

The utilization of volatile fatty acids by PNSB, make them considerable for remediation of areas contaminated with organic pollutants. In the study of Llorens et al. (2012), *R. palustris* which is resistant to aromatic pollutants was also shown to immobilize uranium. 1mM of uranium did not change growth rate or cell concentration of *R. palustris* in stationary phase. Moreover, this bacterium can immobilize U(IV) by reducing it to U(IV)-phosphate or to U(IV)-carboxylate without any loss of viability. The authors suggested that R. palustris, together with a multi-metal ressistant C. metallidurans (can resist up to 30 mM of uranium), can be used for bioremediation of polluted areas containing uranium and other pollutants.

Besides inorganic substances, phytoremediaton of pesticides can also be achieved using PNSB. Pesticides became a concern after their toxicity and toxic effects of various pesticides on various organisms and environments were summarized in the literature (DeLorenzo et al., 2001; Jayaraj et al., 2016; Rajendran, 2016). The PNSB were shown to have pesticide degrading capabilities. For example, a Rhodopseudomonas sp. strain PSB07-8 was found to degrade pesticides fenpropathrin, ypermethrin and bipthenthrin, deltemethrin, cyfluthrin and esfenralerate (Zhang et al. 2014). In a recent study, *Est3385* gene encoding pyrethroid pesticide decomposing esterase enzyme was detected in *R. palustris* PSB-S and proposed to be used in remediation of pesticide residues with its remarkable activity and adaptability (Luo et al. 2018).

Apart from advantages of use of PNSB in agriculture, they can be used in stockbreeding. *R. capsulatus* was shown to be effective in increasing egg quality of hens. Dietary R. capsulatus increased the high-density lipoprotein in the egg and decreased hepatic cholesterol and triglycerides in the laying hens with no negative effects on egg quality and production. The effects were increasing with increased amount of *R. capsulatus*. It was suggested that these bacteria can be used to produce low cholesterol eggs (Salma et al., 2007; Afrose et al., 2010). Likewise, the effect of this bacterium on broiler meat was analyzed. Feeding the broiler chicks with *R. capsulatus*, selenium enriched radish sprouts and vitamin E increased the immunity of the birds and their meat quality (Tsujii et al., 2017). In order to spread the research to other organisms, *R. palustris* and *R. capsulatus* were tried as feed supplement on rats and similar results were obtained in terms of decreased serum cholesterol and triglycerides levels, which reduce the atherosclerosis risk (Tsujii et al. 2007).

REFERENCES

Adiloğlu, A. & Adiloğlu, S. (2005). "Fertilizer Consumption in Turkey Between 1960 and 2002." *Bulgarian Journal of Agricultural Sciences*, *11* (3), 317–21.

Adiloğlu, S. (2016). Using Phytoremediation with Canola to Remove Cobalt from Agricultural Soils. *Polish Journal of Environmental Studies.*, *25*(6), 2251-2254

Adiloğlu, S. (2017). Interaction of Manganese and Some Heavy Metals in Dock (*Rumex patientia* L.) Plant for Remediation of Contaminated Soils. *Desalination and Water Treatment.*, *93*, 335-338.

Adiloğlu, A., Adiloğlu, S. & Karaman, M. R. (2018). "Chemical Fertilizer Consumption of 2000's Years in Tekirdag." In *The International Congress of Eurasian Natural Nutrition and Healthy Life. (DOĞAL 2018)*, *50*, Ankara.

Adiloğlu, S. & Göker, M. (2021). Phytoremediation: Elimination of Hexavalent Chromium Heavy Metal Using Corn (*Zea Mays* L.). *Cereal Research Communications*. Doi: 10.1007/s42976-020-00070-9.

Adiloğlu, S. (2021). Relation of Chelated Iron (EDDHA-Fe) Applications with Iron Accumulation and Some Plant Nutrient Elements in Basil

(*Ocimum Basilicum* L.). *Polish Journal of Environmental Studies., 30*(3).

Afrose, Sadia, Md Sharoare Hossain, Ummay Salma, Abdul Gaffar Miah. & Hirotada Tsujii. (2010). "Dietary Karaya Saponin and Rhodobacter Capsulatus Exert Hypocholesterolemic Effects by Suppression of Hepatic Cholesterol Synthesis and Promotion of Bile Acid Synthesis in Laying Hens." *Cholesterol*, 2010, (June). Hindawi: 272731. doi:10.1155/2010/272731.

Bais, Harsh P., Tiffany L. Weir, Laura G. Perry, Simon Gilroy. & Jorge M. Vivanco. (2006). "The Role of Root Exudates in Rhizosphereinteractions with Plants and Other Organisms." *Annual Review of Plant Biology, 57* (1), 233–66. doi:10.1146/ annurev. arplant.57.032905.105159.

Batool, Kanza, Fatima tuz Zahra. & Yasir Rehman. (2017). "Arsenic-Redox Transformation and Plant Growth Promotion by Purple Nonsulfur Bacteria Rhodopseudomonas Palustris CS2 and *Rhodopseudomonas Faecalis* SS5." *BioMed Research International*, 2017, (March). Hindawi: 1–8. doi:10.1155/2017/6250327.

Berg, Gabriele. (2009). "Plant–microbe Interactions Promoting Plant Growth and Health: Perspectives for Controlled Use of Microorganisms in Agriculture." *Applied Microbiology and Biotechnology, 84* (1), 11–18. doi:10.1007/s00253-009-2092-7.

Bowen, G. D. & Rovira, A. D. (1999). "The Rhizosphere and Its Management To Improve Plant Growth." *Advances in Agronomy, 66* (January). Academic Press, 1–102. doi:10.1016/S0065-2113(08)60425-3.

Çakmakçi, Ramazan, Figen Dönmez, Adil Aydın. & Fikrettin Şahin. (2006). "Growth Promotion of Plants by Plant Growth-Promoting *Rhizobacteria* under Greenhouse and Two Different Field Soil Conditions." *Soil Biology and Biochemistry, 38* (6), Pergamon: 1482–87. doi:10.1016/J.SOILBIO.2005.09.019.

Canbolat, Mustafa Y., Serdar Bilen, Ramazan Çakmakçı, Fikrettin Şahin. & Adil Aydın. (2006). "Effect of Plant Growth-Promoting Bacteria and Soil Compaction on Barley Seedling Growth, Nutrient Uptake,

Soil Properties and *Rhizosphere Microflora*." *Biology and Fertility of Soils*, *42* (4). Springer-Verlag: 350–57. doi:10.1007/s00374-005-0034-9.

Chen, Jian, Guo Xin Sun, Xiao Xue Wang, Víctor De Lorenzo, Barry P. Rosen. & Yong Guan Zhu. (2014). "Volatilization of Arsenic from Polluted Soil by Pseudomonas Putida Engineered for Expression of the ArsM Arsenic(III) S-Adenosine Methyltransferase Gene." *Environmental Science and Technology*, *48* (17), 10337–44. doi:10.1021/es502230b.

Daly, M. J. & Stewart, D. P. C. (1999). "Influence of 'Effective Microorganisms' (EM) on Vegetable Production and Carbon Mineralization–A Preliminary Investigation." *Journal of Sustainable Agriculture*, *14* (2–3). Taylor & Francis Group: 15–25. doi:10.1300/J064v14n02_04.

Delgado-Baquerizo, Manuel, Angela M Oliverio, Tess E Brewer, Alberto Benavent-González, David J Eldridge, Richard D Bardgett, Fernando T Maestre, Brajesh K Singh. & Noah Fierer. (2018). "A Global Atlas of the Dominant Bacteria Found in Soil." *Science (New York, N.Y.)*, *359* (6373). American Association for the Advancement of Science: 320–25. doi:10.1126/science.aap9516.

DeLorenzo, M. E., Scott, G. I. & Ross, P. E. (2001). "Toxicity of Pesticides to Aquatic Microorganisms: A Review." *Environmental Toxicology and Chemistry*, *20* (1), 84–98. http://www.ncbi.nlm.nih.gov/pubmed/11351418.

Deo, Dominic, Ebru Ozgur, Inci Eroglu, Ufuk Gunduz. & Meral Yucel. (2012). "Photofermentative Hydrogen Production in Outdoor Conditions." In *Hydrogen Energy - Challenges and Perspectives*, edited by D Minic. InTech. doi:10.5772/50390.

Elbadry, M. & Kh. Elbanna. (1999). "Response of Four Rice Varieties to Rhodobacter Capsulatus at Seedling Stage." *World Journal of Microbiology and Biotechnology*, *15* (3). Kluwer Academic Publishers: 363–67. doi:10.1023/A:1008923600036.

Elbadry, M., Gamal-Eldin, H. & Kh. Elbanna. (1999). "Effects of Rhodobacter Capsulatus Inoculation in Combination with Graded

Levels of Nitrogen Fertilizer on Growth and Yield of Rice in Pots and Lysimeter Experiments." *World Journal of Microbiology and Biotechnology, 15* (3). Kluwer Academic Publishers: 393–95. doi:10.1023/A:1008958832402.

Evanko, Cynthia R. & David A Dzombak. (1997). "Ground-Water Remediation Technologies Analysis Center Technology Evaluation Report Remediation of Metals-Contaminated Soils and Groundwater." http://www.gwrtac.xn--org-up0a.

Fekiacova, Z., Cornu, S. & Pichat, S. (2015). "Tracing Contamination Sources in Soils with Cu and Zn Isotopic Ratios." *Science of The Total Environment, 517* (June). Elsevier: 96–105. doi:10.1016/J. SCITOTENV.2015.02.046.

Ferron, Pierre. & Jean-Philippe Deguine. (2005). "Crop Protection, Biological Control, Habitat Management and Integrated Farming. A Review." *Agron. Sustain. Dev, 25* (1), 17–24. doi:10.1051/agro:2004050.

Gamal-Eldin, Hosny. & Khaled Elbanna. (2011). "Field Evidence for the Potential of Rhodobacter Capsulatus as Biofertilizer for Flooded Rice." *Current Microbiology, 62* (2). Springer-Verlag: 391–95. doi:10.1007/s00284-010-9719-x.

Ge, Honglian, Zhonghua Liu. & Fuli Zhang. (2017). "Effect of Rhodopseudomonas Palustris G5 on Seedling Growth and Some Physiological and Biochemical Characteristics of Cucumber under Cadmium Stress." *Emirates Journal of Food and Agriculture, 29* (11), 816–21. doi:10.9755/ejfa.2017.v29.i11.1327.

Glick, B. R., Patten, C. L., Holguin, G. & Penrose, D. M. (1999). *Biochemical and Genetic Mechanisms Used by Plant Growth Promoting Bacteria*. Published By Imperial College Press And Distributed By World Scientific Publishing CO. doi:10.1142/p130.

Harasim, Pawel. & Tadeusz Filipek. (2015). "Nickel in the Environment *." *Journal of Elementology, 20* (2), 525–34. doi:10.5601/jelem.2014.19.3.651.

Hotta, Yasushi, Tohru Tanaka, Hideo Takaoka, Yasutomo Takeuchi. & Makoto Konnai. (1997). "New Physiological Effects of 5-

Aminolevulinic Acid in Plants: The Increase of Photosynthesis, Chlorophyll Content, and Plant Growth." *Bioscience, Biotechnology, and Biochemistry*, *61* (12), 2025–28. doi:10.1271/bbb.61.2025.

Hu, Junli, Xiangui Lin, Junhua Wang, Jue Dai, Ruirui Chen, Jiabao Zhang. & Ming Hung Wong. (2011). "Microbial Functional Diversity, Metabolic Quotient, and Invertase Activity of a Sandy Loam Soil as Affected by Long-Term Application of Organic Amendment and Mineral Fertilizer." *Journal of Soils and Sediments*, *11* (2). Springer-Verlag: 271–80. doi:10.1007/s11368-010-0308-1.

Imhoff, Johannes F. (2006). "The Phototrophic Alpha-Proteobacteria." In *The Prokaryotes*, 41–64. New York, NY: Springer New York. doi:10.1007/0-387-30745-1_2.

James, H. & Herman J Gibb. (2006). *Cobalt and Inorganic Cobalt Compounds* (Concise International Chemical Assessment Document 69). Geneva, Switzerland.

Jayaraj, Ravindran, Pankajshan Megha. & Puthur Sreedev. (2016). "Organochlorine Pesticides, Their Toxic Effects on Living Organisms and Their Fate in the Environment." *Interdisciplinary Toxicology*, *9* (3–4). Slovak Toxicology Society: 90–100. doi:10.1515/intox-2016-0012.

Kang, Zhen, Junli Zhang, Jingwen Zhou, Qingsheng Qi, Guocheng Du. & Jian Chen. (2012). "Recent Advances in Microbial Production of δ-Aminolevulinic Acid and Vitamin B12." *Biotechnology Advances*, *30* (6), 1533–42. doi:10.1016/j.biotechadv.2012.04.003.

Kantachote, Duangporn, Tomorn Nunkaew, Thanawan Kantha. & Sumate Chaiprapat. (2016). "Biofertilizers from Rhodopseudomonas Palustris Strains to Enhance Rice Yields and Reduce Methane Emissions." *Applied Soil Ecology*, *100* (April). Elsevier: 154–61. doi:10.1016/J.APSOIL.2015.12.015.

Karadeniz, A., Topcuoğlu, Ş. F. & Inan, S. (2006). "Auxin, Gibberellin, Cytokinin and Abscisic Acid Production in Some Bacteria." *World Journal of Microbiology and Biotechnology*, *22* (10), 1061–64. doi:10.1007/s11274-005-4561-1.

Karaman, Mehmet Rüştü. (2012). *Bitki Besleme*. Edited by Aydın Adiloğlu, R Brohi, A Güneş, A İnal, M Kaplan, V Katkat, A Korkmaz, et al. Ankara: Dumat Ofset.

Kennedy, Max J., Sarah L Reader Lisa M Swierczynski. & Max Kennedy. (1994). "Historical Perspective Preservation Records of Micro-Organisms : Evidence of the Tenacity of Life." *Microbiology, 140*, 251–54. www.microbiologyresearch.org.

Kirpichtchikova, Tatiana A., Alain Manceau, Lorenzo Spadini, Frédéric Panfili, Matthew A. Marcus. & Thierry Jacquet. (2006). "Speciation and Solubility of Heavy Metals in Contaminated Soil Using X-Ray Microfluorescence, EXAFS Spectroscopy, Chemical Extraction, and Thermodynamic Modeling." *Geochimica et Cosmochimica Acta, 70* (9). Pergamon: 2163–90. doi:10.1016/J.GCA.2006.02.006.

Kis, Mariann, Gábor Sipka, Emese Asztalos, Zsolt Rázga. & Péter Maróti. (2015). "Purple Non-Sulfur Photosynthetic Bacteria Monitor Environmental Stresses." *Journal of Photochemistry and Photobiology B: Biology, 151* (October), 110–17. doi:10.1016/ j.jphotobiol. 2015.07.017.

Koh, Rae-Hyun. & Hong-Gyu Song. (2007). "Effects of Applicaiton of Rhodopseudomonas Sp. on Seed Germination and Growth of Tomato under Axenic Conditions." *Journal of Microbiology and Biotechnology*, *17* (11), 1805–10.

Koku, Harun, Inci Eroglu, Ufuk Gündüz, Meral Yücel. & Lemi Türker. (2002). "Aspects of the Metabolism of Hydrogen Production by Rhodobacter Sphaeroides." *International Journal of Hydrogen Energy, 27* (11–12), 1315–29. doi:10.1016/S0360-3199(02)00127-1.

Lee, Kang-Hyeong, Rae-Hyun Koh. & Hong-Gyu Song. (2008). "Enhancement of Growth and Yield of Tomato by Rhodopseudomonas Sp. under Greenhouse Conditions." *The Journal of Microbiology, 46* (6). The Microbiological Society of Korea: 641–46. doi:10.1007/s12275-008-0159-2.

Llorens, Isabelle, Guillaume Untereiner, Danielle Jaillard, Barbara Gouget, Virginie Chapon. & Marie Carriere. (2012). "Uranium Interaction with Two Multi-Resistant Environmental Bacteria: Cupriavidus

Metallidurans CH34 and Rhodopseudomonas Palustris." Edited by Vipul Bansal. *PLoS ONE*, *7* (12). Public Library of Science: e51783. doi:10.1371/journal.pone.0051783.

Luo, Xiangwen, Deyong Zhang, Xuguo Zhou, Jiao Du, Songbai Zhang. & Yong Liu. (2018). "Cloning and Characterization of a Pyrethroid Pesticide Decomposing Esterase Gene, Est3385, from Rhodopseudomonas Palustris PSB-S." *Scientific Reports*, *8* (1). Nature Publishing Group: 7384. doi:10.1038/s41598-018-25734-9.

Madigan, Michael T. & Deborah O Jung. (2009). "An Overview of Purple Bacteria: Systematics, Physiology, and Habitats." *The Purple Phototrophic Bacteria*, *28*, 1–15. doi:10.1007/978-1-4020-8815-5.

McLaughlin, M. J., Hamon, R. E., McLaren, R. G., Speir, T. W. & Rogers, S. L. (2000). "Review: A Bioavailability-Based Rationale for Controlling Metal and Metalloid Contamination of Agricultural Land in Australia and New Zealand." *Australian Journal of Soil Research*, *38* (6). CSIRO PUBLISHING: 1037. doi:10.1071/SR99128.

Meng, Xiang Yan, Jie Qin, Li Hong Wang, Gui Lan Duan, Guo Xin Sun, Hui Lan Wu, Cheng Cai Chu, Hong Qing Ling, Barry P. Rosen. & Yong Guan Zhu. (2011). "Arsenic Biotransformation and Volatilization in Transgenic Rice." *New Phytologist*, *191* (1), 49–56. doi:10.1111/j.1469-8137.2011.03743.x.

Nagata, Satoshi, Keiko Yamaji, Nobuhiko Nomura. & Hiroshi Ishimoto. (2015). "Root Endophytes Enhance Stress-Tolerance of *Cicuta Virosa* L. Growing in a Mining Pond of Eastern Japan." *Plant Species Biology*, *30* (2), 116–25. doi:10.1111/1442-1984.12039.

Nair, Smitha, Archana Joshi-Saha, Sudhir Singh, Ramachandran V., Surya Singh, Vidya Thorat, Kaushik, C. P., Susan Eapen. & D'Souza, S. F. (2012). "Evaluation of Transgenic Tobacco Plants Expressing a Bacterial Co–Ni Transporter for Acquisition of Cobalt." *Journal of Biotechnology*, *161* (4). Elsevier: 422–28. doi:10.1016/J.JBIOTEC.2012.07.191.

Ncube, Lindani, Pearson Nyari S Mnkeni. & Marc Olivier Brutsch. (2016). *Agronomic Suitability of Effective Micro- Organisms for Tomato*

Production Agronomic Suitability of Effective Micro-Organisms for Tomato Production, no. February 2011. doi:10.5897/AJAR10.515.

Nookongbut, Phitthaya, Duangporn Kantachote, Kannan Krishnan. & Mallavarapu Megharaj. (2017). "Arsenic Resistance Genes of As-Resistant Purple Nonsulfur Bacteria Isolated from As-Contaminated Sites for Bioremediation Application." *Journal of Basic Microbiology, 57* (4), 316–24. doi:10.1002/jobm.201600584.

Nunkaew, Tomorn, Duangporn Kantachote, Hiroshi Kanzaki, Teruhiko Nitoda. & Raymond J. Ritchie. (2014). "Effects of 5-Aminolevulinic Acid (ALA)-Containing Supernatants from Selected Rhodopseudomonas Palustris Strains on Rice Growth under NaCl Stress, with Mediating Effects on Chlorophyll, Photosynthetic Electron Transport and Antioxidative Enzymes." *Electronic Journal of Biotechnology, 17* (1). Elsevier: 19–26. doi:10.1016/ J.EJBT.2013. 12.004.

Nunkaew, Tomorn, Duangporn Kantachote, Teruhiko Nitoda. & Hiroshi Kanzaki. (2015). "Selection of Salt Tolerant Purple Nonsulfur Bacteria Producing 5-Aminolevulinic Acid (ALA) and Reducing Methane Emissions from Microbial Rice Straw Degradation." *Applied Soil Ecology, 86*, 113–20. doi:10.1016/j.apsoil.2014.10.005.

Rajendran, Soorya. (2016). "Acute Toxic Effect of Pesticides on Human Beings: A Review." *Research & Reviews: Journal of Chemistry, 5* (2). Research and Reviews: 87–94. http://www.rroij.com/open-access/acute-toxic-effect-of-pesticides-on-human-beings-a-review-.php?aid=77974.

Ramanjaneyulu, A. V. & Gajendra Giri. (2006). "Phytoremediation - A Review." *Agricultural Reviews, 27* (3), 216–22.

Sakpirom, Jakkapan, Duangporn Kantachote, Tomorn Nunkaew. & Eakalak Khan. (2017). "Characterizations of Purple Non-Sulfur Bacteria Isolated from Paddy Fields, and Identification of Strains with Potential for Plant Growth-Promotion, Greenhouse Gas Mitigation and Heavy Metal Bioremediation." *Research in Microbiology, 168* (3). Elsevier Masson: 266–75. doi:10.1016/J.RESMIC.2016.12.001.

Salma, U., Miah, A. G., Tareq, K. M. A., Maki, T. & Tsujii, H. (2007). "Effect of Dietary Rhodobacter Capsulatus on Egg-Yolk Cholesterol and Laying Hen Performance." *Poultry Science*, *86* (4), 714–19. doi:10.1093/ps/86.4.714.

Sasaki, Ken, Masanori Watanabe, Yoshito Suda, Akihiro Ishizuka. & Napavarn Noparatnaraporn. (2005). "Applications of Photosynthetic Bacteria for Medical Fields." *Journal of Bioscience and Bioengineering*, *100* (5), 481–88. doi:10.1263/jbb.100.481.

Sheraz Mahdi, S., Hassan, G. I., Samoon, S. A., Rather, H. A., Dar, A. S. & Zehra, B. (2010). "Bio-Fertilizers in Organic Agriculture." *Journal of Phytology*, *2* (10), 42–54.

Siles, José A. & Rosa Margesin. (2016). "Abundance and Diversity of Bacterial, Archaeal, and Fungal Communities Along an Altitudinal Gradient in Alpine Forest Soils: What Are the Driving Factors?" *Microbial Ecology*, *72* (1). Springer: 207–20. doi:10.1007/s00248-016-0748-2.

Simmons, Shaneka S., Raphael D. Isokpehi, Shyretha D. Brown, Donee L. Mcallister, Charnia C. Hall, Wanaki M. Mcduffy, Tamara L. Medley., et al. (2011). "Functional Annotation Analytics of *Rhodopseudomonas Palustris* Genomes." *Bioinformatics and Biology Insights*, *5*, (January), BBI.S7316. doi:10.4137/BBI.S7316.

Sojka, G. A. (1978). "Metabolism of Nonaromatic Organic Compounds." In *The Photosynthetic Bacteria*, edited by Roderick K. Clayton and WR Sistrom, 707–18. New York: Plenum Press.

Su, Pin, Xinqiu Tan, Chenggang Li, Deyong Zhang, Ju'e Cheng, Songbai Zhang, Xuguo Zhou., et al. (2017). "Photosynthetic Bacterium Rhodopseudomonas Palustris GJ-22 Induces Systemic Resistance against Viruses." *Microbial Biotechnology*, *10* (3). Wiley-Blackwell: 612–24. doi:10.1111/1751-7915.12704.

Tok, Hasan Hayri. (1998). *Toprak Biyolojisi*. Tekirdağ: Trakya Üniversitesi Tekirdağ Ziraat Fakültesi.

Tsujii, Hirotada, Abdul Gaffar Miah, I Takeda. & Ummay Salma. (2017). "Dietary Effect of Selenium-Enriched Radish Sprouts, Vitamin E, and Rhodobacter Capsulatus on Hypocholesterolemia and Immunity of

Broiler." *Gorgan University Of Agricultural Sciences & Natural Resources, 5* (1). Gorgan University Of Agricultural Sciences & Natural Resources, 71–81. doi:10.22069/PSJ.2017.12411.1233.

Tsujii, Hirotada, Masahiro Nishioka, Ummay Salma, Abdul G. Miah, Takaaki Maki. & Myung G. Lee. (2007). "Comparative Study on Hypocholesterolemic Effect of Rhodopseudomonas Palustris and Rhodobacter Capsulatus on Rats Fed a High Cholesterol Diet." *Animal Science Journal, 78* (5). Wiley/Blackwell (10.1111), 535–40. doi:10.1111/j.1740-0929.2007.00473.x.

van der Heijden, Marcel, G. A., Richard D. Bardgett. & Nico M. van Straalen. (2008). "The Unseen Majority: Soil Microbes as Drivers of Plant Diversity and Productivity in Terrestrial Ecosystems." *Ecology Letters, 11* (3), 296–310. doi:10.1111/j.1461-0248.2007.01139.x.

Verbon, Eline H. & Louisa M. Liberman. (2016). "Beneficial Microbes Affect Endogenous Mechanisms Controlling Root Development." *Trends in Plant Science, 21* (3), 218–29. doi:10.1016/j.tplants.2016.01.013.

Weaver, P. F., Wall, J. D. & Gest, H. (1975). "Characterization of Rhodopseudomonas Capsulata." *Archives of Microbiology, 105* (3), 207–16. doi:10.1007/BF00447139.

Wong, Wai-Tak, Ching-han Tseng, shu-hua hsu, huu-sheng Lur, Chia-Wei Mo, Chu-ning huang, shu-Chiung hsu, kung-Ta Lee. & Chi-Te Liu. (2014). "Promoting Effects of a Single Rhodopseudomonas Palustris Inoculant on Plant Growth by Brassica Rapa Chinensis under Low Fertilizer Input." *Microbes Environ, 29* (3), 303–13. doi:10.1264/jsme2.ME14056.

Xu, Hui-Lian, Ran Wang. & Md. Amin U. Mridha. (2001). "Effects of Organic Fertilizers and a Microbial Inoculant on Leaf Photosynthesis and Fruit Yield and Quality of Tomato Plants." *Journal of Crop Production, 3* (1). Taylor & Francis Group: 173–82. doi:10.1300/J144v03n01_15.

Xu, Jiangbing, Youzhi Feng, Yanling Wang, Xiaosan Luo, Jianwu Tang. & Xiangui Lin. (2016). "The Foliar Spray of Rhodopseudomonas Palustris Grown under Stevia Residue Extract Promotes Plant Growth via Changing Soil Microbial Community." *Journal of Soils and Sediments*, *16* (3), 916–23. doi:10.1007/s11368-015-1269-1.

Yang, Zhimin, Zuoliang Chang, Lihong Sun, Jingjin Yu. & Bingru Huang. (2014). "Physiological and Metabolic Effects of 5-Aminolevulinic Acid for Mitigating Salinity Stress in Creeping Bentgrass." *PloS One*, *9* (12). Public Library of Science: e116283. doi:10.1371/journal.pone.0116283.

Zhang, Songbai, Chunhui Zhu, Yong Liu, Deyong Zhang, Xiangwen Luo, Feixue Cheng, Ju', N. A., Cheng, E. & Yuanhua Luo. (2014). "Biodegradation of Fenpropathrin by a Novel Rhodopseudomonas Sp. Strain PSB07-8." *International Journal of Environmental Engineering*, *6* (1), 55. doi:10.1504/IJEE.2014.057833.

Zhou, Yang, Honghui Zhu, Shenglei Fu. & Qing Yao. (2017). "Variation in Soil Microbial Community Structure Associated with Different Legume Species Is Greater than That Associated with Different Grass Species." *Frontiers in Microbiology*, *8*. Frontiers Media SA: 1007. doi:10.3389/fmicb.2017.01007.

In: The Future of Phytoremediation
Editor: James C. Flores
ISBN: 978-1-53619-625-2

Chapter 7

PHYTOREMEDIATION OF HEAVY METALS CONTAMINATED RICE FIELDS USING WILD-FOUND PLANTS: A NEW APPROACH FOR SUSTAINABLE AGRICULTURE

Nuril Hidayati * ***and Dwi Setyo Rini***
Research Center for Biology- Indonesian Institute of Sciences
Jl. Raya Jakart, West Java, Indonesia

ABSTRACT

Heavy metals pollution in agricultural lands is becoming a serious problem since it has major impact on reducing agricultural productions in quantity and quality. Meanwhile conventional efforts for remediation have not been widely implemented due to high-cost constraints. A sustainable cost-effective technology that can be applied in contaminated sites is urgently needed. Phytoremediation known as plant-base technology could be one alternative. This project aims to assess potential Pb and Cd accumulators and possible application for phytoremediation in

* Corresponding Author's E-mail: nuril2019@yahoo.com.

the contaminated rice fields. Lead (Pb) and cadmium (Cd) are among the most commonly found contaminants in agricultural areas. Six wild-found plants that are predominantly grown in contaminated area were examined in an in-situ trials in Pb and Cd contaminated rice fields. Determining variables, namely biomass production, Pb and Cd uptake, Bioconcentration Factor (BCF) that indicate the ability of plants to remove metals from the soil and Translocation Factor (TF) that indicate the ability of the compound to be transferred from plant roots to shoot were examined. There are six plant species namely *Colocasia* sp. *Ipomoea fistulosa*, *Eichhornia crassipes*, *Hymenachne amplexicaulis*, *Saccharum spontaneum* and *Acorus calamus* that showed considerably high Pb and Cd accumulation after 9 weeks of growing period. The highest (BCF) for root Pb was 0,671 found in *S. spontaneum* and the highest BCF of shoot Pb was 0,558 in *I. fistulose*. Whereas the highest BCF for Cd in the root was 0,566 found in *S. spontaneum* and the highest BCF for Cd in the shoot was 0,475 in *S. spontaneum*. Translocation Factor for Pb is highest at 0,904 in *I. fistulosa* and 0,828 in *S. spontaneum*. Meanwhile the highest TF for Cd was 0,927 observed in *A. calamus* and 0,839 in *S. spontaneum*. Regarding variables of biomass production and Pb and Cd accumulation levels in the plants it can be concluded that *E. crassipes*, *H. amplexicaulis* and *A. calamus* can be suggested as potential accumulator candidates for phytoremediation of contaminated rice fields.

Keywords: accumulators, cadmium, contamination, lead, rice field, phytoremediation

1. Introduction

Heavy metals contamination in soils and waters is a common problem at many hazardous waste sites. Once released into the soil and waters, most heavy metals are strongly retained and their effects can last for a long time and they are present at the elevated concentration at many National Priority List (NPL) sites. Heavy metals are toxic to people and pose a great risk for ground water and agriculture irrigation (Squire, 2001).

Heavy metals contamination in agricultural lands becoming a serious problem since this causes declining in agriculture production in quantity and quality and thus food security and food safety. About 75% of agricultural lands in Indonesia are categorized as critical lands where soil fertility continues to decline (Sidik, 2011). Several factors that cause the declining quality of agricultural land are the continuous use of agrochemical fertilizers and pesticides in high doses and the contamination of the agricultural field by heavy metals pollutants sourced from industrial wastes disposed in the river flow which then is used as a source of water irrigation. Several cases of agricultural land contamination in Indonesia are also caused by heavy metals contamination from mining and industrial. It was reported that many pollutants found in paddy fields such as Pb and Cd. Approximately 21- 40% of the 106 thousand ha of paddy fields in the West Java Coastal Area are polluted by heavy metals, where 4 – 7% of them are categorized heavily polluted by Pb (Kasno et al., 2000; Miseri et al., 2000).

Meanwhile conventional efforts for remediation of the contaminated agricultural lands have not been widely implemented due to high-cost constraints. In line with the SDG's goals, to achieve a better and more sustainable future, a sustainable approach for cleaning up polluted agricultural lands for sustainable agriculture is urgently needed. One of the easy, cos-effective and sustainable technology that can be applied is phytoremediation. The concept of phytoremediation basically relies on the species richness and a greater number of species represent high remediation efficiency of the pollutants present in the environment (Upadhyay et al., 2020). Phytoremediation technique is based on the fact that plant roots have the ability to extract and accumulate heavy metals in considerably high quantity (Wislocka, 2006). Detoxification of contaminants such as heavy metals with plants seems to be a feasible option for cleaning soils and waters. Some cases show that it is highly possible of using plants for remediation), especially in the countries that are rich in biodiversity prospect as hyperaccumulators (Rahmansyah et al., 2009; Prasat & Frietas, 2003; Rodriguez et al., 2007; Yu et al., 2005).

The plant types that are of particular interest for use as remediation are those that absorb, store and detoxify contaminants (Gerth, 2000). Hyperaccumulators are plant species that can both accumulate and tolerate shoot concentrations of heavy metals to level 100 times greater than those normally observed (Ebbs et al., 2000). Hyperaccumulators are plants that can accumulate specific trace elements to concentrations in their shoots far higher than other plants growing on the same substrates (Reeves et al., 2018). The hyperaccumulation threshold values are set at: 100 μg g-1 for Cd, Se and Tl; 300 μg g-1 for Co, Cr and Cu; 1000 μg g-1 for As, Ni and Pb; 3000 μg g-1 for Zn; and 10 000 μg g-1 for Mn (van der Ent et al., 2013a). Some wild-found plant species are proven to accumulate considerably high concentration of heavy metals such as *Hymenachne amplexicaulis* accumulated 65,35 ppm of Pb and 2,27 ppm of Cd, *Colocasia sp.* accumulated 41,7 ppm of Pb and up to 11,66 ppm of Cd, *Panicum paludosum* accumulated 40,20 ppm Pb and 5,10 ppm Pb, *Cyperus sp* accumulated 38,5 ppm Pb and 4,6 ppm Cd (Rahmansyah et al., 2009; Juhaeti, et al., 2009b; Hidayati & Rini, 2020).

All above-mentioned data demonstrated the ability of hyperaccumulators of Pb and Cd in contaminated soils and waters. In situ technique is popular and often used a technique to remedy the soil contaminated with heavy metals (He et al., 2015; Khan et al., 2015). In spite of this, a very few field trials were reported for phytoremediation of lead (Pb) and cadmium (Cd) metals in agriculture lands. A study was conducted to investigate the role of some metal hyperaccumulators in remediation of agricultural soils. Since plants vary in their ability to accumulate specific contaminants, it is necessary to select plant species that can accumulate and tolerate the contaminants at a given site and to analyze the capacity of their potency as hyperaccumulator. Wild-found plants are assumed to be the most suitable candidates for accumulators since they have already been adapted with the environment.

Sustainable technology Innovations could be used as the basis for development policy towards the protection of agricultural areas, serving both food security function and strengthening sustainability. There still has not been any research explaining sustainable technology innovations approach for overcoming degradation and pollution in agriculture lands to support sustainable development programmes in this region. This project was performed to assess the potency of several wild-found plants as heavy metals, lead (Pb) and Cadmium (Cd) bioaccumulators on an *in-situ* trials of contaminated rice fields and to examine the possible solution for cleaning up contaminated rice fields as one alternative of sustainable approach in innovation technology.

2. Materials and Methods

2.1. Study Area

The site of the study was determined based on the facts that the site, namely Bekasi Regency has been known as one of the most important rice production center in West Java with total rice field area of 42.771 ha in 1918. Secondly, most of agriculture areas in Bekasi was found to be polluted by heavy metals due to agro-chemical residue and industrial waste that becoming so excessive that threatening sustainable development in agriculture.

Bekasi Regency belongs to West Java Province, Indonesia with Latitude of -6,2467° or 6° 14' 48" south; Longitude of 107,1083° or 107° 6' 30" east with surrounding neighbors: East Jakarta, Karawang and North Jakarta (Figure 1). The experimental plot was located in Sukaindah Village, Sukakarya Sub District, Bekasi (S: 06° 06' 52,20;" E: 107° 09' 31,68") that representing 2400 ha - 2500 ha of rice fields. The plot location was determined based on the following criteria: (i) Close to rivers and irrigation which are polluted by a high level of Pb and Cd. In this case, the location of the *in-situ* field is located close to Citarum River; (ii) Agricultural fields are exposed by a high level of Pb and Cd; (iii) Accessible for supervision and maintenance; (iv) Representative to total the rice field area.

The status of Pb and Cd in the experimental plot are as follows: (i) Irrigation water from Cikarang river I: Pb status 7,824 ppm and Cd 0,592 ppm; (ii) Irrigation water from Cikarang river II: Pb status 10,275 and Cd 0,665 ppm; (iii) Rice field in study site: the level of Pb was 10,26 ppm, Cd was 0,98 ppm. Meanwhile, the value of the standard quality of Pb in water is 0,08-2,0 ppm, in soils is 1,0-100 ppm and in rice is 2,0 ppm. The quality standard of Cd in water is 0,05-0,1 ppm, in soils is 0,1-10 ppm and in rice is 0,24 ppm (Hidayati and Rini, 2018). Therefore, the status of Pb and Cd levels in the testing plots were categorised above quality standard.

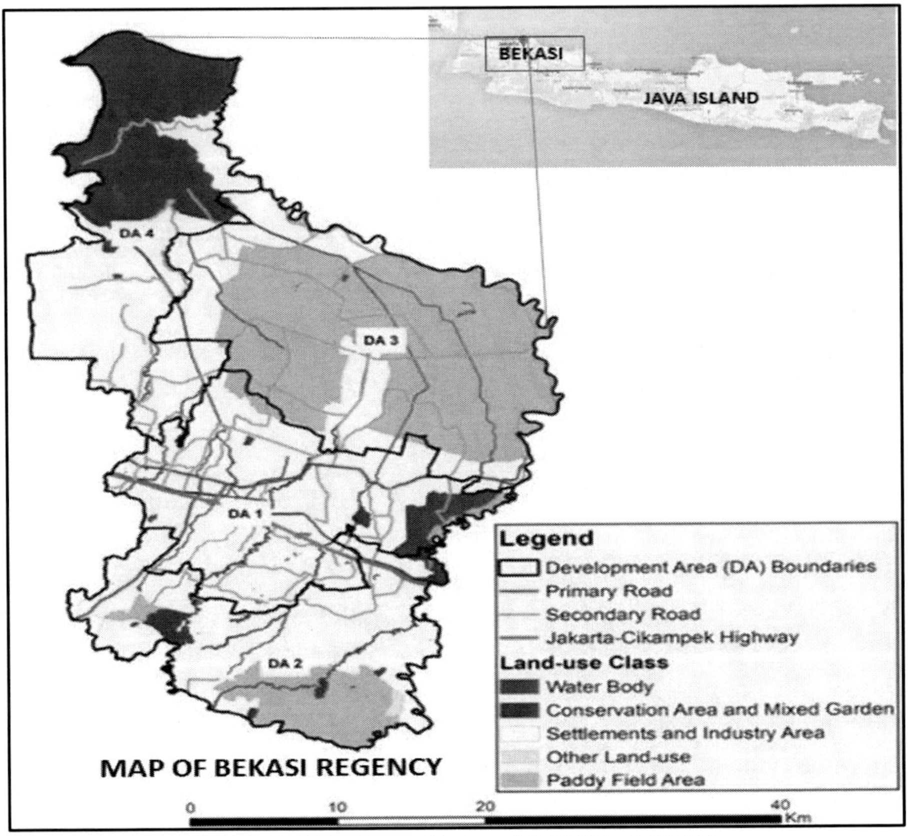

Sustainability 2020, 12(19), 8094; https://doi.org/10.3390/su12198094; http://www.worldeasyguides.com/asia/indonesia/bekasi/where-is-bekasi-on-map-java/.

Figure 1. Position of Bekasi Regency in Java Island, Indonesia.

2.2. Experimental Procedure

The trials used six wild-found plants that are commondly found in the area to promote an easy and cheap implementation. Six plants that are predominantly growing in the area were chosen to minimize the risk of experimental failure due to the poor plant growth and adaptability. Variables of agronomic traits and plant metals uptake from the soils were observed to determine the effectiveness of metal extraction by plant accumulator candidates.

Six candidates of Pb and Cd accumulator plants based on the previous *ex-situ* selection were tested, namely 1) *Colocasia* sp. (talas liar); 2) *Ipomoea fistulosa* Mart. Ex Choisy. (tablo); 3) *Eichhornia crassipes* (Mart.) Solms. (eceng gondok); 4) *Hymenachne amplexicaulis* (Rudge) Nees.(rumput paitan); 5) *Saccharum spontaneum* L. (kaso) and 6) *Acorus calamus* L.(rumput manggala). The basic standard dosage of inorganic fertilizer was applied to optimize the harvested biomass. The experiment was designed in a randomized block consisting of six types of accumulator plants with three replications. The plants were subjected to heavy metal contamination in the *in-situ* field for 9 weeks starting on 17 August 2018 up to 23 October 2018.

Determining Variables observed were the agronomic traits (i.e., adaptability, plant growth rate and harvested shoot and root biomass). Plant growth rate was calculated as the increment of biomass per week. Total biomass production was measured as root and shoot biomass harvested after 9 weeks, as a given time interval for all of the plant. Variable of heavy metals status was measured as heavy metals concentrations in the soil, before planting and after harvesting and heavy metals concentration in the plant shoots and roots before planting and after harvesting.

Sampling was performed twice, before planting and after harvesting (after 9 weeks). Samples which consist of roots and shoots biomass were collected from three individuals of plant in each treatments as replications. The plant samples (separated roots and shoots) were thoroughly washed by distilled water to remove all adhering soil particles. Samples were then cut into small pieces and oven dried to constant weights at 80 °C. Each dried sample was ground to powder using a wearing blender. Soil samples were collected from each treatment plot as a composites with three replications. In order to determine total levels of heavy metals, the composite soil samples were air-dried to constant weights and ground into fine powder for analysis. Approximately 5g for each sample (roots, shoots and soils) from each replication was provided for Pb and Cd analysis. Plant botanical identification was conducted in Herbarium Bogoriense, Research Center for Biology, Indonesian Institute of Sciences as well as the *ex-situ* and molecular selection that have been conducted prior to this *in-situ* study. Analysis of Pb and Cd accumulation was done in Research Center for Biology, Indonesian Institute of Sciences, and Chemistry Laboratory of IPB University, Bogor, Indonesia.

2.2.1. Data Analysis

Heavy metal concentrations data from soils and plants (shoots and roots) were used to calculate the bioconcentration and translocation factors. Phytoextraction potential can be estimated by calculation of bioconcentration factor (or biological absorption coefficient) and translocation factor. Bioconcentration factor (BCF) is commonly used to indicate the ability of plants to remove metal compounds from the soil. Bioconcentration factor (BCF), which is defined as the ratio of the total concentration of element in the harvested plant tissue (C plant) to its concentration in the soil in which the plant grows (C) (Mellen et al., 2012), was calculated as follows:

$$\text{BCF} = \frac{\text{Concentration in plant organ}}{\text{Concentration in sediment}}$$

Meanwhile, translocation factor (TF) is a variable that is commonly used to indicate the ability of the compound to be transferred from plant roots to other organs/shoot (Mellen et al., 2012). Parameters of BCF and TF were analysed to assess whether plants could be categorized as accumulators which were calculated as follows:

$$\text{TF} = \frac{\text{BCF of shoot}}{\text{BCF of root}}$$

Data on plant growth, biomass production, heavy metals accumulation as well as bioconcentration factors and translocation factor, were analysed using SAS programme and Windows Excel and are presented in table and chart format.

3. Results and Discussion

3.1. Plant Growth

Plant growth is one indicator to understand the adaptability of tested plants to environment, in this case adaptability to rice field environment and adaptability to Pb and Cd contaminated soils. The six plants tested showed significant growth differences since they had different characteristics and responses to the environmental conditions. Agronomic traits that was presented in Figure 2 and 3 and Table 2 indicated that four out of six accumulator plants subjected to Pb and Cd contamination have performed high adaptability and biomass production, except for S. spontaneum and I. fistulosa which were not in accordance with the growing conditions in the rice fields. The six plants showed significant growth differences since they have different characteristics and responses to the environmental conditions. E. crassipes, A. calamus and H. amplexicaulis are suitable with the rice field environment therefore these three species performed the highest rates of growth and biomass production. After 9 weeks growing period, the highest growth rate and biomass production were observed on E. crassipes, which totally increased 221,33 kg with the growth rate of 24,59 kg/week, followed by A. calamus which totally increased 128,46 kg with the rate of 14,27 kg/week. H.amplexicaulis as rank 1 accumulator candidate that is highly expected in terms of potential accumulation of Pb and Cd, performed total biomass production of 51,93 kg with the growth rate of 5,77 kg / week. While Colocasia sp., I. fistulosa and S. spontaneum produced respectively of 17,28 kg, 2,15 kg and 2,13 kg biomass (Figure 2 and Figure 3 and Table 2). Total plant biomass increased by $\pm$ 10 times in average. This indicated that some plants have considerably hight adaptability to Pb and Cd contamination and therefore can be considered as potential accumulators with regards to the biomass production. Among the six plants tested, S. spontaneum and I. fistulosa showed the most inferior performances, therefore, these two plants were not suitable to the in-situ rice field trials. S. spontaneum and I. fistulosa belong to terrestrial plants that require a better drainage for their growing environments.

The results of the present research accordance to some other findings that wild-found plant species found to be more adaptable to polluted environment and could be recommended for phytoremediation like *C. demersum* (Chen et al., 2015; Dogan et al. 2018), *Lemna minor* and *Lemna gibba* (Sasmaz et al., 2016) and *C. demersum* that is highly effective aquatic plants used for phytoremediation against different pollutants found in water bodies (Hassan and Al-Khalidi, 2018). The adaptability is proven to be related with the genetic and physiological characteristics of each plant species (Nkumar 2018). Different researchers highlighted the variable response of plants exposed to different heavy metals at variable concentrations (Duman and Koca, 2014). Heavy metal concentrations and time exposure interval significantly affected the plant growth which in turn resulted in variable response on fresh and dry wt of phytoremediated plants. Both fresh and dry wt parameters reflect the toxicity level of phytoremediated plants (Aasim et al., 2020).

Since the success of the phytoremediation technique is determined by increasing plant biomass and accumulation of heavy metals in accumulator plants (Karger et al., 2015). Therefore some remarks need to be considered in the phytoremediation practices in order to achieve the optimum results. Appropriate agronomic practices need to be applied, based on insights from laboratory and field tests, to maximize yields of the selected accumulators (Nkumar, 2018). Local plant species are highly recommended as accumulators due to their adaptation to local climate conditions. Some studies also suggested of using wild species as heavy metals accumulator since they have high potential to be used for phytoremediaion (Cˇurlik et al., 2016; Mudarrisna and Siahaan, 2014). In addition, to improve plant growth and metal uptake it can be employed combination of inorganic fertilizer and compost as a chelator such as *Agrobacterium sp.* (Rostiati, 2018).

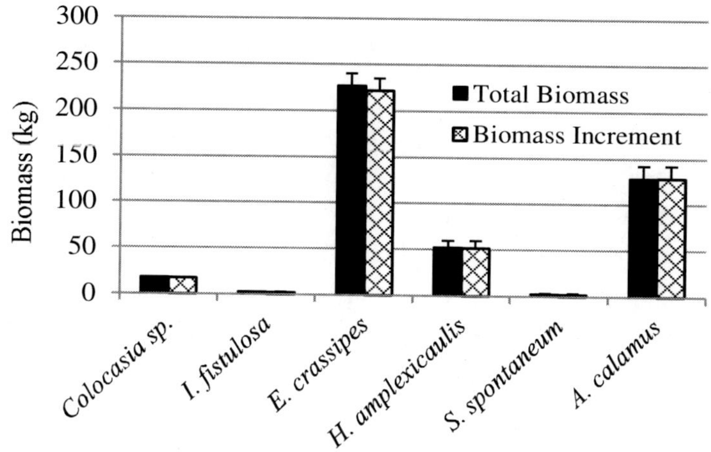

Figure 2. Plant Biomass Increment After 9 Weeks.

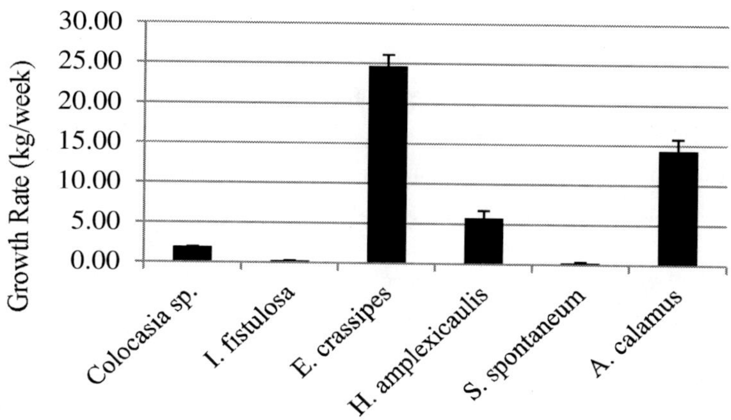

Figure 3. Plant Growth Rate in 9 weeks growing period.

3.2. Heavy Metals Uptake

Assessment on heavy metal accumulation in the tested accumulator plants after 9 weeks of growing period indicated that there are five out of six potential accumulator plant were suggested in terms of Pb and Cd accumulation levels in the plant shoot and root. The highest increase of Pb accumulation in plant shoot was found in *S spontaneum* (3,517 ppm) followed by *Colocasia* sp. (3,239 ppm), I. *fistulosa* (2,961 ppm), *A. calamus* of 2,321 ppm and *H. amplexicaulis* (2,043 ppm). Similar pattern also found in Pb increase in root that the highest was occurred on *S spontaneum* (4,635 ppm) followed by *Colocasia* sp.(4,396 ppm), *E. crassipes* (3,900 ppm), *A. calamus* (3,857 ppm) and *H. amplexicaulis* (3,425 ppm) (Figure 4). Meanwhile the increase of Cd in plant shoot after 9 weeks was observed that the highest level occurred in *S. spontaneum* (0,339 ppm) followed by *A. calamus* (0,306), *Colocasia* sp. (0,304 ppm), *E. crassipes* of 0,299 ppm and *H. amplexicaulis* (0,258 ppm). Whereas the highest level of Cd in the plant root was observed on *S.spontaneum* (0,434 ppm), followed by I. *fistulosa* (0,428 ppm), *H. amplexicaulis* (0,419 ppm), *E. crassipes* of 0,415 ppm and *Colocasia* sp. (0,353 ppm) (Figure 5).

Levels of Pb and Cd accumulation in the plants within 9 weeks was considerable high, which were 1-3 ppm Pb and 0,1-0,4 ppm Cd respectively. There were no significant differences in Pb uptake between plant species, although the real data show that several plants tend to accumule higher Pb than others. Total of heavy metal accumulated by plants is determined not only by the accumulation concentration in plants but also by the amount of biomass production, thus the accumulation capacity of heavy metals is a function of the concentration of metals accumulated in plants and total of harvested biomass. Regarding these two factors, total amount of Pb and Cd uptake were then become much higher within 9 weeks, especially for *E. crassipes*, *A. calamus* and *H. amplexicaulis* which experienced biomass increase up to 10 times. It can be concluded that regarding variables of biomass production and Pb and Cd concentration in the plants, *E. crassipes*, *H. amplexicaulis* and *A. calamus* can be suggested as the expected accumulator candidates. In line with this findings, Singh et al., (2017) reported that some wild plants can be accumulator for some heavy metals. Among aquatic plant, *Panicum antidotale* is a hyperaccumulator for one or other metals up to six metals (Cd, Co, Fe, Mn, Pb, and Zn). Among terrestrial plants, *Lantana camara* and *Ageratum conyzoides* are hyperaccumulator for Cd, Cu, Fe, Mn, Pb, and Zn. It is evident that all these plant species have potential to remediate various inorganic pollutants and can act as bioaccumulator.

It was observed that Pb and Cd accumulation were found to be higher the roots than that in the shoot. Corresponds to the present findings, many other findings also suggest that heavy metal in roots is generally larger than canopy (stems and leaves). The predominantly low TF values indicate also that in most cases the heavy metals are accumulated in the roots, which is known in many other scientific studies (Parzych, 2016). Only in a few rare cases do they migrate to shoots, e.g., Mn and Zn in all species (Andras et al., 2016). Heavy metals are commonly found in roots and leaves. The pattern of metals accumulation, including Pb and Cd was found to be determined by some factors such as pH, temperature, Cd concentration in media, and even concentration of elements other than Cd (Mahajan & Kaushal, 2018), soil fertility, the presence of microorganism, availability of water, and the presence of ligants and chelating agents (Rosariastuti et al., 2018).

Metals removal through phytoremediation emerges as a sustainable technology for contaminated soil as well as wastewater. Phytoremediation has high performance results when compared with other conventional technologies for metal removal. The plant species from wide group of families have been recognized as Cd and Pb hyperaccumulators. Different hyperaccumulating plants have varied abilities to accumulate, sequester, and detoxify Cd and Pb at physiological and molecular level. Despite the lot of developments made in field of Cd and Pb phytoremediation from contaminated soil and water, only a limited number of research studies have taken place in field conditions. Hence, there is an urgent need for research on improving experimental of phytoremediation soil and water (Mahajan & Kaushal, 2018).

Although the levels of Pb and Cd increased significantly in the plant during the testing period, they can not be correlated with the decrease of Pb and Cd levels in the soils of testing plots (Table 1). In this study case however, the Pb and Cd levels in the soil before and after harvesting cannot be used as a benchmark for the effectiveness of phytoextraction. This is because the system applied in the experimental plot is not a closed system. The water regulation in the experimental plots must be adjusted to the water regulation for the surrounding agricultural areas. Therefore this system still be exposed by Pb and Cd from the flowing-in contaminated irigation water continuously as all of the rice field in the area are also irrigated from. Similar *in-situ* test and results was performed that the use of wastewater as a source of continuous irrigation can accumulate heavy metals significantly (Ma et al., 2015).

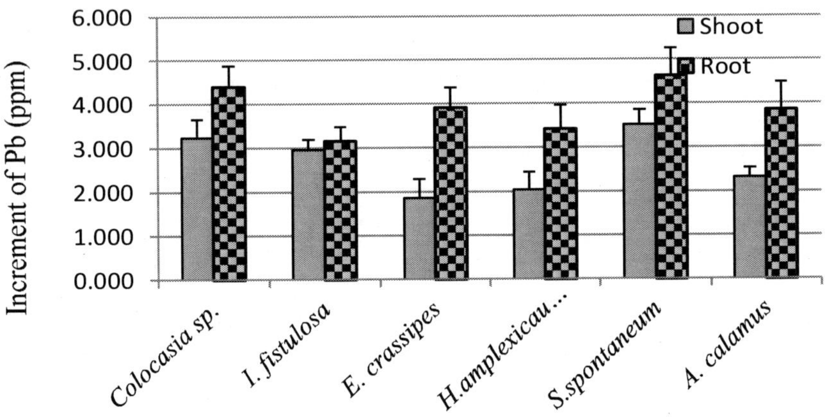

Figure 4. Increment of Pb in Plants After 9 Weeks in Bekasi.

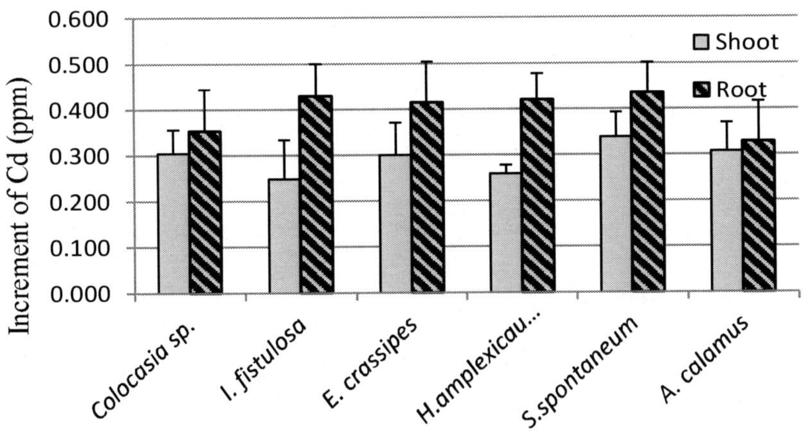

Figure 5. Increment of Cd in Plants After 9 Weeks in Bekasi.

3.3. Bioconcentration Factor and Translocation Factor

Table 1 and 2 represent the value of BCF and TF of tested plants. The overall root BCF is greater than canopy BCF for both Pb and Cd. There was no significant differences in BCF values amongs plant species. The real data showed that the highest BCF for root Pb was 0,671 in *S. spontaneum* and the highest BCF of Pb in *I. fistulose* was 0,558. Whereas the highest BCF for Cd at the root was found in *S. spontaneum* at 0,566 and BCF for Cd at the highest canopy of 0,475 in *S. spontaneum.* There were significant differences in the value of root TF. Translocation Factor (TF) for Pb is highest at 0,904 in *I. fistulose* and 0,828 in *S. spontaneum,* While TF for Cd is highest at 0,927 in *A. calamus* and 0,839 in *S. spontaneum.* Although the value of TF could not reach 1, however some accumulator candidates tend to reach higher values of TF and potentially become accumulators, namely *I. fistulose and A. calamus.*

Plant showed variation in lead concentration, bioconcentration and translocation factors. This results were supported by some other findings. A significant and better uptake of Cd was observed in the hydroponic system as compared to soil cultures as Bio Concentration Factor (BCF) and Translocation Factor (TF) were more than 1 but on high exposure of Cd (Mahajan and Kaushal, 2018). The highest translocation factors for all metals, Pb, Cu, and Zn, were found for root-stem translocation, with values of 1,09; 2,80, and 2,12 respectively (Takarina & Pin, 2017). Plants that have bioconcentration and translocation factors > 1 can be used as bioaccumulators. Bioconcentration values > 2 are considered to be high values (Mellen et al., 2012). Plants can be used as phytostabilizers if they have bioconcentration factors > 1 and translocation factors < 1 and as phytoextractors if they have bio-concentration factors < 1 and translocation factors > 1 (Usman et al., 2013). In a recent field study with Ganges ecotype of *T. caerulescen*, the role of soil geochemical factors and plant-soil interactions for Cd uptake are highlighted by hyperaccumulating plants. Those studies imply the importance of understanding site-specificity heavy metals containment and soil properties to be considered before phytoremediation of actual field sites (Mahajan and Kaushal, 2018), since the transfer and accumulation of the heavy metals into the plant organs is usually substantially influenced by various factors. The factors responsible for remediation of Cd and Pb by plants are pH, temperature, metals concentration in media, and even concentration of elements pas well as by heavy-metal content in soil and by soil characteristics (e.g., the mould and clay minerals content), soil reaction, cation exchange capacity, or by Ca = Mg content ratio (Cˇ urlík et al., 2015; Mahajan and Kaushal, 2018). The effect of pH on the metal mobility is a variable but very important factor. Mobility of metals in soils with low pH decreases in the order Cd > Ni > Zn >Mn> Cu > Pb (Vamerali et al., 2010). The metal concentration in plants increases as a function of metal concentration in the soil. The BCF value of an accumulator plant varies depending on the existing heavy metals in the contaminated area (Smitha and Sugandha, 2019).

The translocation factors for heavy metals in plants should be more than one in order to be considered as bioaccumulators (Siahaan et al., 2013). The process of heavy metal binding in plants occurs via the mechanisms of heavy metal accumulation and tolerance. Plant roots can thus contribute to the prevention of heavy metal accumulation in the surrounding environment; high concentrations of heavy metal contaminants in sediments may be stored/extracted by the plants (Takarina & Pin, 2017). The metal concentration in plants increased as a function of metal concentration in the soil. BCF values varied from site to site, indicating variation in heavy contamination of the soil due to release of wastewater from nearby industries. High BCF values of the plant species indicate poor quality of river water. Increase in BCF values is found linear with initial soil metal concentrations. This exhibits plant as bioindicators activity for soil contamination (Smita & Sugandha, 2019).

Table 1. Bio-Concentration Factor (BCF) and Translocation Factor (TF) of Accumulator Plants in Bekasi District

Bekasi		**Pb in Plant (ppm)**					**Cd in Plant (ppm)**				
Accumulators	Samples	Pb Soil	Before Planting	After 9 Weeks	TF shoot/ root	BCF	Cd Soil	Before Planting	After 9 Weeks	TF shoot/root	BCF
Colocasia sp.	Shoot	8,835	0,562	3,801	0,723	0,430	0,796	0,036	0,340	0,808	0,427
	Root	8,835	0,858	5,254		0,595	0,796	0,068	0,421		0,529
I. fistulosa	Shoot	6,972	0,926	3,887	0,904	0,558	0,983	0,072	0,319	0,624	0,325
	Root	6,972	1,144	4,302		0,617	0,983	0,083	0,511		0,520
E. crassipes	Shoot	9,108	1,042	2,906	0,620	0,319	0,847	0,078	0,377	0,795	0,445
	Root	9,108	0,785	4,685		0,514	0,847	0,059	0,474		0,560
H.amplexicaulis	Shoot	10,264	1,033	3,076	0,687	0,300	0,942	0,077	0,335	0,701	0,356
	Root	10,264	0,953	4,378		0,436	0,942	0,059	0,478		0,507
S.spontaneum	Shoot	7,718	0,774	4,291	0,828	0,556	0,845	0,062	0,401	0,839	0,475
	Root	7,718	0,546	5,181		0,671	0,845	0,044	0,478		0,566
A. calamus	Shoot	10,033	0,882	3,203	0,656	0,319	0,836	0,075	0,381	0,927	0,456
	Root	10,033	1,027	4,884		0,487	0,836	0,084	0,411		0,492

Table 2. Statistical analysis of plant growth rate, bioconcentration factor (BCF) and translocation factor (TF)

Plant species	Plant growth rate (kg/week)	BCF for Pb	TF fo Pb	BCF for Cd	TF for Cd
Colocasia sp.	2,50c	0,595a	0,723a	0,529a	0,808a
I. fistulosa	0,61d	0,617a	0,904a	0,520a	0,624ab
E. crassipes	24,55a	0,514a	0,620ab	0,560a	0,795a
H.amplexicaulis	5,90c	0,436a	0,687ab	0,507a	0,701a
S.spontaneum	0,50d	0,671a	0,828a	0,566a	0,829a
A. calamus	14,50b	0,481a	0,656ab	0,492a	0,927a

Note: Different alphabetical notation behind the values indicate according analysis (ANOVA p 0.05), the values are significantly different.

CONCLUSION

Our study suggests that some plants are proven capable of extracting and accumulating heavy metals and can be effectively used as bioaccumulators in phytoremediation techniques to clean up heavy metals contamination in agricultural areas. Some plants can produce high biomass and absorb high contaminants while other plants cannot, implying that plants respond differently to different environmental conditions. There are six plant species namely Colocasia sp. Ipomoea fistulosa Mart. ex Choisy, [Eichhornia crassipes (Mart.) Solms], [Hymenachne amplexicaulis (Rudge) Nees], Saccharum spontaneum L. and Acorus calamus L. that showed considerably high Pb and Cd accumulation. Regarding variables of biomass production and Pb and Cd concentration levels in the plants, it can be concluded that E. crassipes, H. amplexicaulis and A. calamus can be suggested as the expected accumulator candidates for phytoremediation of contaminated rice fields. Despite the data suggested a positive and promising prospect for the implementation, yet continuous research is required to obtain appropriate and suitable techniques for effective results.

ATTACHMENT

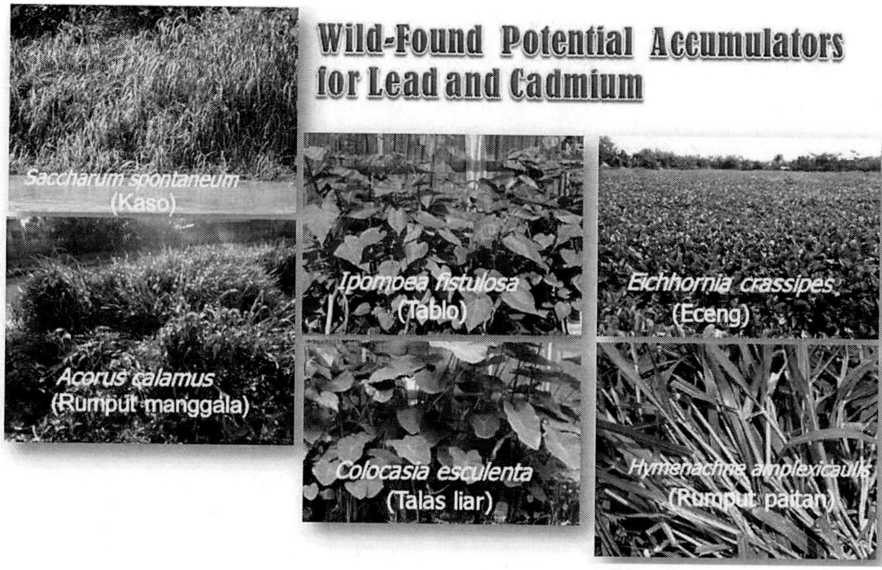

Figure 6. Wild-found Potential Accumulator Plants for Lead (Pb) and Cadmium (Cd)

ACKNOWLEDGMENTS

Appreciation and many thanks was addressed to SMARD - Indonesia Ministry of Agriculture and Research Center for Biology - Indonesian Institute of Sciences (LIPI). Many thanks and appreciation was also addressed to some institutions and individuals in Bekasi who gave contributions to this project, namely UPTD Cilamaya Kulon, Karawang Agriculture Agency (Dinas Pertanian Kabupaten Karawang), H. Adun and others.

REFERENCES

Aasim M, S Aydin, M Karatas, M E Aydin, C Sogukpinar and C Sevinc. 2020. Bioaccumulation potential of *In vitro* regenerated plants of *Ceratophyllum demersum* against Chromium - A lab study. *Asian J. gric. Biol. 8*(3): 233-239. DOI:https://doi.org/10.35495/ajab.2019.11.516.

Andráš P, I Turisová, G Buccheri, J M X Matos and V Dirner. 2016. Comparison of heavy-metal bioaccumulation properties in *Pinus* sp. and *Quercus* sp. in selected European Cu deposits. *Web Ecol.*, 16, 81–87, 2016 www.web-ecol.net/16/81/2016/doi:10.5194/we-16-81-2016 © Author(s) 2016. CC Attribution 3.0 License.

Chen M, L L Zhang, J Li, X J He and J C Cai. 2015. Bioaccumulation and tolerance characteristiof a submerged plant (*Ceratophyllum demersum* L.) exposed to toxic metal lead. Ecotoxicol. *Environ. Saf.* 122: 313–321.

Cˇ urlík J, M D Kolesár, O Durža and E Hiller. 2015. Dandelion (*Taraxacum officinale*) and Agrimony (*Agrimonia eupatoria*) as Indicators of Geogenic Contamination of Flysch Soils in Eastern Slovakia, *Arch. Environ. Con. Tox.*, Epub 9 August 2015, doi:10.1007/s00244-015-0206-z, 2015.

Dogan M, M Karatas and M Aasim. 2018. Cadmium and lead bioaccumulation potentials of an aquatic macrophyte *Ceratophyllum demersum L.: A laboratory study. Ecotoxicology and Environmental Safety.* 148: 431-440.

Duman F and F D Koca. 2014. Single and combined effects of exposure concentration and duration on biological responses of *Ceratophyllum demersum*L. Exposed to Cr species. *Int. J. Phytoremediat.* 16: 1192–1208.

Ebbs S, L Kochian, M Lasat, N Pence, T Jiang. 2000. An integrated investigation of the phytoremediation of heavy metal and radionuclide contaminated soils: From laboratory to the field. Di dalam:Wise D L, Trantolo DJ, Cichon E J, Inyang H I, Stottmeister U (eds.).

Bioremediation of Cotaminated Soils Marcek Dekker Inc New York Basel p 745-769.

Gerth A 2000 Phytoremediation of soil and sludge with special examination of heavy metal ontamination In: Wise D L, Trantolo D J, Cichon E J, Inyang H I, Stottmeister U (ed) : *Bioremediation of Cotaminated Soils*. Marcek Dekker Inc. New York Basel P 787-809.

Hassan N A and S H H Al-Khalidi. 2018. Phytoremediation of lead by *Ceratophyllum demersum* - Lab Work. *Int. J. Curr. Microbiol. Appl. Sci*. 7(5): 2002-2008.

He F, J Gao, E Pierce, P J Strong, H Wang and L Liang. 2015. In situ remediation technologies for mercury-contaminated soil. *Environmental Science and Pollution Research* 22(11): 8124-8147.

Hidayati N & D S Rini. 2018. Possible solution of cleaning up heavy metals contamination in agriculture area using accumulator plants. *Proceeding The 7th Iternational Symposium for Sustainable Humanorphere (ISSH) "Advance in Science and Technology for the Future of Sustainable Humanosphere."* ISSN: 2579-5554. P 83-92.

Hidayati N & D S Rini. 2020. Evaluation of lead and cadmium accumulators for phytoremediation of contaminated rice fields in Karawang, West Java. *BIODIVERSITAS* Vol.21 no. 5 May 2020. ISSN: 1412-033X. E-ISSN: 2085-4722. DOI: 10.13057/biodiv/d2105xx.

Juhaeti T, N Hidayati dan F Syarif. 2009b. Uji potensi tumbuhan akumulator merkuri untuk fitoremediasi lingkungan tercemar akibat kegiatan penambangan emas tanpa izin (PETI) di kampong Leuwi Bolang, Desa bantar Karet, Kecamatan Nanggung, Bogor. *Jurnal Biologi Indonesia* 6 (1): 1-12. [Indonesian].

Karger M, O G Clark, W H Hendershot, P Jutras, S O Prasher. 2015. Immobilization of trace metals in contaminated urban soil amended with compost and biochar. *Water, Air Soil Pollution* 226: 191.

Kasno A, J S Adiningsih, Sulaeman, Subowo. 2000. *Pollution status of Pb and Cd in intensified paddy fields in the northern coast of West Java. Proceedings of the VII National Congress of the Indonesian Soil Science Association* p. 1537-1546.

Khan F, M J Khan, A Samad, Y Noor, M Rashid and B Jan. 2015. In-situ stabilization of heavy metals in agriculture soils irrigated with untreated wastewater. *Journal of Geochemical Exploration* 159: 1-7.

Mahajan P and J Kaushal. 2018. *Review Article* Role of Phytoremediation in Reducing Cadmium Toxicity in Soil and Water. *Hindawi Journal of Toxicology* Volume 2018, Article ID 4864365, 16 pages. https://doi.org/ 10.1155/2018/4864365.

Ma S, H Zhanga, S Ma, R Wang, G Wang, Y Shao and C Li. 2015. Ecotoxicology and Environmental Safety Effects of mine wastewater irrigation on activities of soil enzymes and physiological properties, heavy metal uptake and grain yield in winter wheat. *Ecotoxicology and Environmental Safety* 113:483-490.

Mellem J J, H Baijnath, B Odhav. 2012. Bioaccumulation of Cr, Hg, As, Pb, Cu and Ni with the ability for hyperaccumulation by *Amaranthus dubius*. *African Journal of Agricultural Research*. 7(4): 591-596, doi: 10.5897/AJAR11.1486.

Miseri R A, A Z P B Santoso, I Novianto. 2000. Impact of motor vehicle fumes on lead levels (Pb) in soil and plants around the Cirebon Palimanan highway. *Proceedings of the VII National Congress of the Indonesian Soil Science Association* p 1457-1466.

Muddarisna N & BC Siahaan. 2014. Application of organic matter to enhance phytoremediation of mercury contaminated soils using local plant species: a case study on small-scale gold mining locations in Banyuwangi of East Java. *J. Degraded and Mining Lands Management.* 2 (1): 251-258. DOI: 10.15243/jdmlm.2014.021.251.

Nkrumah P N, R L Chaney, J L Morel. 2018. Agronomy of 'metal crops' used in agromining. Springer International Publishing AG 2018 A. van der Ent et al., (eds) *Agromining: farming for metals, mineral resource reviews,* DOI 10.1007/978-3-319-61899-9_2. January 2018. St.Lucia, QLD, Australia.

Parzych A E. 2016. Accumulation of chemical elements by organs of *Sparganium erectum* L. and their potential use in phytoremediation process, *J. Environ. Eng.*, 17, 89–100, 2016.

Prasat, M N V, H M O Frietas 2003 Metal hyperaccumulation in plants-Biodiversity prospecting for phytoremediation technology *Electronic J. Biotechnol.* Vol 6.

Rahmansyah M, N Hidayati dan T Juhaeti (Eds.). 2009. *Tumbuhan Akumulator untuk Fitoremediasi Lingkungan Tercemar Merkuri dan Sianida Penambangan Emas.* LIPI Press. ISBN 978-979-799-423-5. 143 hal.

Reeves R D, A J M Baker, T Jaffré, PD Erskine, G Echevarria, A van der Ent. 2018. A global database for plants that hyperaccumulate metal and metalloid trace elements. *New Phytologist,* pp. 218: 407–411.

Rodriguez L, J Rincon, I Asencio, C L Rodriguez. 2007. Capability of selected crop plants for shoot mercury accumulation from polluted soils: Phytoremediation perspectives. *Int. J. Phytorem.* 9: 1-13.

Rosariastuti R, Barokah U, Purwanto, Supriyadi. 2018. Phytoremediation of Pb contaminated rice field using combination of *Agrobacterium sp.* I3, compost and ramie (*Boehmeria nivea*). *J. of Degraded and Mining ands Management.* 5(4) 1381-1388 DOI: 10.15243/jdmlm.2018.054.1381.

Sasmaz M, E Obek and A Sasmaz. 2016. Bioaccumulation of uranium and thorium by *Lemna minor* and *Lemna gibba* in Pb-Zn-Ag tailing water. *Bull. Environ. Contam. Toxicol.* 97(6):832-837.

Siahaan M T A, Ambariyanto, B Yulianto. 2013. Pengaruh pemberian timbal (Pb) dengankonsentrasi berbeda terhadap klorofil, kandungan timbal pada akar dan daun, serta struktur histologi jaringan akar anakan mangrove *Rhizophora sp.mucronata. Journal of Marine Research.* 2(2): 111119. [The effect of lead (Pb) with different concentrations on chlorophyll, lead content in roots and leaves, and the histological structure of the root tissue of Rhizophora sp.mucronata mangrove saplings. *Journal of Marine Research*.]

Sidik J M. 2011. Camkan ini, 75 persen lahan pertanian kritis. Sangat kecil sekali tanah produktif kita. *Antara News. www.antarnews.com* (22 Pebruary 2016).

Singh N, M Kaur and J K Katnoria. 2017, Analysis on bioaccumulation of metals in aquatic environment of Beas River Basin: A case study from Kanjli wetland, *GeoHealth*, 1, 93–105, doi:10.1002/2017GH000062.

Smita C V and S Sugandha. 2019. Determination of Bioconcentration factor (BCF) for copper and zinc in *Pongamia pinnata* plant *and Blumea malcolmii* plant signifying their role as bioindicators of soil contamination *Research Journal of Chemistry and Environment* Vol. 23 (4) April (2019) *Res. J. Chem. Environ.*

Squire V R. 2001. Soil pollution and remediation: issues, progress and prospects. Di dalam: *Prosiding Workshop Vegetation Recovery in Degraded land Areas*. Kalgoorlie, Western Australia, 27 October.

Takarina N D & T G Pin. 2017. Bioconcentration Factor (BCF) and Translocation Factor (TF) of Heavy Metals in Mangrove Trees of Blanakan Fish Farm. *Makara Journal of Science*, 21/2 (2017), 77-81doi: 10.7454/mss.v21i2.7308.

Upadhyay A K, R Singh, and D P Singh. 2020. Phytoremediation and Sustainable Developmental Policies and Practices in: A. K. Upadhyay et al. (eds.), *Restoration of Wetland Ecosystem: A Trajectory Towards a Sustainable Environment*, https://doi.org/10.1007/978 981 13 7665 8_2. ISBN 978-981-13-7664-1. ISBN 978-981-13-7665-8 (eBook). 248 p.

Usman A R A, R S Alkredaa and M I Al-Wabel. 2013. Heavy metal contamination in sediments and mangroves from the coast of Red Sea: *vicennia sp. marina* as potential metal bioaccumulator. *Ecotoxicol. Environ. Saf.* 97: 263-270, http://dx.doi.org/10.1016/j.ecoenv.2013.08.009.

Vamerali T, M Bandiera and G Mosca. 2010. Field crops for phytoremediation of metal- contaminated land. A review, *Environ. Chem. Lett.*, 8, 1–17, 2010.

van der Ent A, A J M Baker, R D Reeves, A J Pollard, H Schat. 2013a. Hyperaccumulators of metal and metalloid trace elements: Facts and fiction. *Plant and Soil* 362: 319–334.

Wislocka W, J Krawczyk, A Klink, L Morrison. 2006 Bioaccumulation of Heavy Metals by Selected Plant Species from Uranium Mining Dumps in the Sudety Mts., *Poland. Polish J. of Environ. Stud.* 15(5): 811-818.

Yu X, P Zhou, X Zhou, Y Liu. 2005. Cyanide removal by Chinese vegetation-quantification of the Michaelis-Menten kinetics. *Environ. Sci. Pollut. Res. Int.* 12: 221-6.

In: The Future of Phytoremediation
Editor: James C. Flores
ISBN: 978-1-53619-625-2

Chapter 8

PHYTOREMEDIATION: A SCIENTOMETRIC ANALYSIS OF CONTRIBUTION IN SCIELO AND SCOPUS DATABASES

Ísis Danielle Sousa[1], Marcelino Benvindo Souza[1], Nayane Cristina Pires[2], Jackellyne Bruna Sousa[3], Liliane Santos Camargos[2] and Lucas Anjos Souza[1,§]

[1]Instituto Federal Goiano – IF Goiano,
Campus Rio Verde. Rio Verde, GO, Brasil
[2]Universidade Estadual Paulista – UNESP,
Campus 2 Ilha Solteira. Ilha Solteira, SP, Brasil
[3]Escola Superior de Agricultura "Luiz de Queiroz" –
ESALQ – USP. Piracicaba, SP, Brasil

ABSTRACT

The Preservation of soils is a key determinant for human survival. Soil biodiversity influences the quantity and quality of food consumed, so it is a valuable natural resource. However, soil biodiversity is being

[§] Corresponding Author's E-mail: lucas.anjos@ifgoiano.edu.br.

affected by global man-made changes and currently, after incorrect practices in its use, there are warning signs of soil contamination with diverse pollutants. The hazardous nature of wastes can be mitigated by the application of conventional decontamination methods, such as incineration, composting and landfilling, although such techniques are expansive. A possibility is to consider a viable and promising alternative technique for treatment of contaminated soils by applying biological techniques that have as main objective the stimulation of the growth of the microorganisms using the contaminants as source of energy and food, creating a propitious environment for remediation. It is the case of Phytoremediation, a technique that uses plants and associated soil microbes to decrease the concentration or toxic effects of contaminants in environments, specially in soil. However, it is necessary to know the potential of plant around the world and, Brazil, which holds one of the greatest natural resources and biodiversity on the planet, we lack a ruge knowledge of plants with phytoremediation potential. In the last 20 years, many discussions have been raised, in both academic and business environment, converging to the same consensus: there is a delay in innovations from natural products and derivatives derived from our rich biodiversity. Therefore, this scientometric research bring a broad view of what have been carried in Brazil and abroad concerning phytoremediation studies associated with a critical view about what should be done to boos this research field in order to generate more ecofriendly restoration process to recover polluted lands.

Keywords: contamination, bioeconomics, bioremediation, environment

INTRODUCTION

The earth's crust is composed of a cover layer called soil, which suffers impacts of various types, direct and indirect damage caused by anthropic exploration of the environment and among which can be mentioned the spillage of chemical products thus damaging several life forms. One of the main aims of soil functions is to provide sufficient and primordial natural resources for the perpetuation and surviving of plants, animals and the human race and these functions depend on the soil balance and relationships between physical, chemical and biological properties, as well as on its structure and composition (Heiderscheidt, 2016).

According to Brasil, 12305/2010, tailings are solid residues that, after all the possibilities of treatment and recovery by available and economically viable technological processes have been exhausted, do not present any other possibility than the environmentally appropriate final disposal. The waste, on the other hand, is all material, substance, object or discarded good resulting from human activities in Society which the final destination of proceeds, proposes to proceed or is obliged to proceed, in solid or semi-solid states.

This waste is generated by activities such as smelting, mining, manufacturing, traffic emissions, urban waste, use of agricultural fertilizers or pesticides and industrial effluents (Morgan 2013; Chibuike & Obiora 2014) and has increased dramatically in the last decades, causing contamination of the soil by heavy metals like Ni, Zn, Pb and Cu (Chibuike & Obiora 2014) and, it can be considered as generalized (Al nagger et al., 2013). Soil degradation caused by heavy metals has opposite and very damaging effects on the environment and ecosystem across the planet (Li et al. 2013; Chen et al. 2015), as these trace elements are not metabolizable and, therefore, the degradation process is impossible leading then they to accumulate in soils and, consequently, in plants (Ghoneim et al. 2014).

Regarding the application of phytoremediation, there is a need to identify plants with high tolerance to trace elements, that is, plants capable of growing and developing without suffering from toxic effects (Souza et al. 2013). By definition, phytoremediation is a technique based on the strategy of removing toxic compounds from the environment by using plants. Because it is economically viable, it provides less impact on the environment (Ali et al. 2013) and, the choice of plant species determines the efficiency of the technique. In such approach, the trace elements can be accumulated in roots or transported to aerial tissues such as stems and leaves (Moosavi & Seghatoleslami 2013). However, studies related to the tolerance and distribution pattern of trace elements in the body of different plant species are necessary.

Brazil is known for presenting the greatest biological diversity in the world in its different biomes. However, these natural resources are neglected by the frantic expansion of exploratory economic activities or even overestimated by conservationist movements (Pimentel et al. 2015). For this reason, the country has been losing scientific space for knowledge of its own biodiversity because it considers the predominant economic interests, while in reality, Brazil could take advantage of the resources that nature offers combined with sustainable development.

In this context, through scientometric evaluation, we evaluate the contribution of Brazilian research compared to international research in the area of phytoremediation to identify the advancement of knowledge in this area in different parts of the world and, thus, making possible to suggest ways of fostering the development of this area of research.

Materials and Methods

Bibliographic Data Compilation

We carried out a bibliographic survey in two databases: Scielo (Scientific Electronic Library Online) and Scopus by using the indexing term in the Scielo database "*phytoremediation*" in "*All indexes*", in the Scopus database we use the term "*phytoremediation*" in "*Article, title, abstract and key words*". The search was carried out comprising the periods between 2000 and 2020 in the two databases.

Grouping and Categorizing Data

Distribution of the Number of Articles Published per Year

With the publication data, we organize the articles by year of publication, thus generating a time trend graph of studies on phytoremediation in the last 20 years for both databases.

Grouping of Articles Published by Journal

In order to be aware about journals that publish articles about phytoremediation area, we organize the articles retrieved by scientific journal, regardless of the year of publication.

Grouping of Data by Type of Contaminant

In order to assess the general classification of the most studied contaminants, we categorized the articles according to the type of contaminant addressed in the study, whether organic or inorganic. We also distribute articles according to the type of contaminant per journal.

Location of Publications

To identify the level of contribution, by nation, in research about phytoremediation, we carried out an assessment of geographical distribution, by continent, of the number of countries that published with the theme phytoremediation taking into account the host institution of the first author of the study.

Classification of Identified Journals

From the grouping of articles according to the type of pollutant studied, organic or inorganic, we classified the journal by the Impact Factor (IF).

Grouping of Species Worked by Biome

Using the data from the Scielo database, we detail all the species studied in the researches retrieved from this database and, according to them, we searched for the biome of predominant occurrence in Brazil.

Data Analysis

A Pearson correlation ($P < 0.05$) was performed with the number of publications by the year of publication, in order to verify whether there was a significant increase in the number of studies over the years. Another Pearson correlation analysis was performed with the phytoremediation aspects (inorganic or organic) in Brazilian studies. In order to know the geographical aspects of the countries in scientific collaboration (institutional link of the first author), the distribution of localities and continents by presenting international data is presented.

Results and Discussion

Scientific Trend on Phytoremediation in Brazil

Studies concerning phytoremediation and/or trace elements tolerance may be within basic or applied research and there is a difference in the supporting resources to finance researches and, therefore, there may be an impact on the quantity and/or quality of researches published within a study field. In the research of Brauna et al. (2017), searching the topic "*sustainable remediation*", in the Scielo database, it was retrieved 196 publications in the range between the years 1980 and 2016. The authors identified that the first publications occurred in 1999 and from the 2000s onwards, with an interest in incorporating sustainability in decision-making processes regarding the remediation of contaminated areas, through the dissemination of the term "*sustainable remediation*", reflecting a perception that remediation activities can bring environmental impacts, social and economic, both positive and negative.

The number of researches published about phytoremediation in the Scielo database was low, in the early 2000s with only 2 publications, compared to 2010, with 20 total researches being 15 with inorganic pollutants and 5 with organic pollutants. At 2011, it reached a total of 26 publications, being 18 with inorganic pollutants and 8 with organic pollutants (Figure 1). After this period of maximum production, there was a decrease in production and, in 2013, we observed that only 14 published researches were related to phytoremediation. Considering the study range time, from 2000 to 2020, the Brazilian journals that most published researches about phytoremediation were: "Planta Daninha" and "Revista Brasileira de Ciência do Solo", with a total of 31 and 22 works, respectively (Table 1). The analysis of the historical series shows that there was an increase in publications between 2000 and 2020.

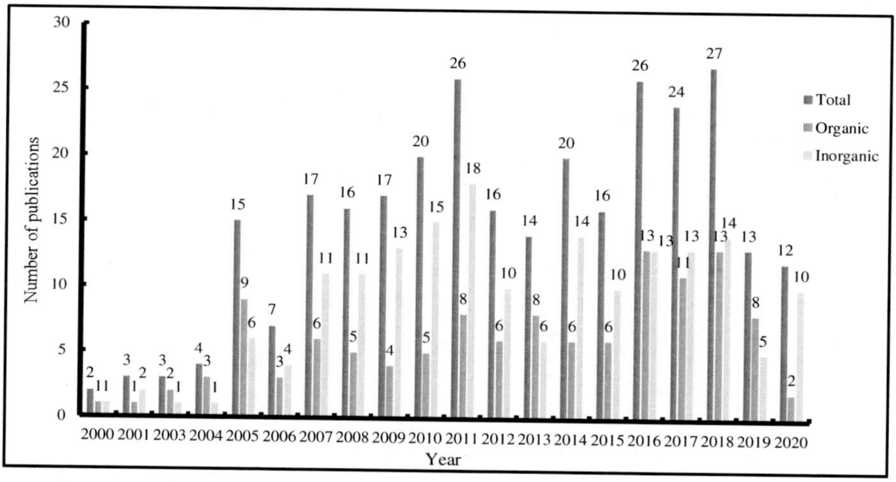

Figure 1. Total number and depiction of organic and inorganic pollutants researches published in journals indexed in Scielo database from 2000 to 2020.

In the range tima studied by Brauna et al. (2017), they found a gradual increase in publications over the years, and between 1999 and 2008 there was a stable performance that from 2009, the total researches published raised, reaching the peak in 2008, with 42 researches, around to 21.5% of total researches. Such evidence indicates the increase in studies regarding phytoremediation published in journals indexed in the Scielo database, which may represent an increase in the interest of the scientific research community.

In our study, we carried out a survey of journals indexed to the Scielo database that published studies about phytoremediation and, in the period from 2000 to 2020, we found that the journals "Planta Daninha" and "Revista Brasileira de Ciência do Solo" published a total of 31 and 22 works, respectively (Table 1). The journal "Planta Daninha", published more studies related to inorganic pollutants (26 studies), and the journal "Revista Brasileira de Ciência do Solo", published more studies regarding organic pollutants (18 studies). Therfore, it is possible to observe that the journals has audiences active in different areas of study and, their studies, can be complementary considering the resolution of environmental problems caused by inorganic and organic pollutants.

We observed that journals in the area of plant physiology also publish researches related to phytoremediation, although this area of study is also of interest to ecology, environmental engineering and environmental chemistry. Silva (2016) observed that phytoremediation researches are multidisciplinary and require knowledge of areas such as microbiology, engineering, ecology, geology and chemistry. Therefore, such field of study is increasingly being sought, studied and performed by a large number of authors, as this technique explores various fields of knowledge and becomes vastly simple and accessible to contextualize with these areas.

Table 1. Total number of researches published in the first 15 journals indexed in Scielo database as well as the number of researches with organic and inorganic pollutants from 2000 to 2020

Journals	Total	Organic	Inorganic
Planta Daninha	31	5	26
Revista Brasileira de Ciência do Solo	22	18	4
Revista Brasileira de Engenharia Agrícola e Ambiental	13	8	5
Scientia Agricola	12	10	2
Brazilian Journal of Plant Physiology	11	9	2
Journal of soil science and plant nutrition	9	8	1
Brazilian Journal of Biology	7	6	1
Revista Árvore	7	5	2
Bragantia	6	4	2
Ciência e Agrotecnologia	5	4	1
Ciência Rural	5	3	2
Electronic Journal of Biotechnology	5	4	1
Revista Ciência Agronômica	5	3	2
Revista de Biologia Tropical	5	3	2

Due to the increase in researches about phytoremediation in journals related to agrarian sciences, it is possible to consider that this area has an interest in sustainable technology for the treatment of contaminated areas through clean technologies such as phytoremediation, as observed in the trend of publications of the period studied (Figure 1). Therefore, it is important to consider the need to increase the budget of research funding to generate or optimize environmental processes related to phytoremediation.

Classification of Journals that Publishes about Phytoremediation

It is important to know the reaching potential of the researches carried out, therefore, we've considered important to bring the impact factors of journals that published researches in phytoremediation area and, therefore, we categorize the journals based on their impact factor.

Table 2. Number of studies with organic and inorganic pollutants in scientific journals indexed at Scielo data base and organized according to impact factors

	Pollutant	
IF	**Organic**	**Inorganic**
0-0.99	105	119
1 – 1.99	14	51
2 – 2.99	1	8
3 – 3.99	0	0
4 – 4.99	0	0
5- 5.99	0	0
6 – 6.99	0	0

The impact factor of the researches published about Phytoremediation in the Scielo database, three groups were identified: in the range 0 to 0.99, we identified 105 and 119 studies for organic and inorganic, respectively; in the range between 1 and 1.99, we identified 14 and 51 studies on organic and inorganic pollutants, respectively; and, finally, in the range of 2 to 2.99, we identified 1 and 8 studies on organic and inorganic pollutants, respectively. Considering impact factor higher than 3, we did not observe publications (Table 2). These observations suggest that most of the studies carried out and published in Scielo database, may not be contributing to the breakdown or establishment of new paradigms within the phytoremediation area, since we have not observed publications in journals with higher impact factors.

Identification of Botanical Groups Studied in Researches Published in Scientific Journals Indexed in Scielo Database

From searching the Scielo database and, after categorizing the results in studies that studied organic and inorganic contaminants, we also identified which botanical groups were the most studied that were published in journals indexed in Scielo databased. The most representative botanical group for studies considering tolerance to toxic metals was herbaceous plants (82 studies), followed by herbaceous extract, aquatic macrophyte, tree plants, shrubs and arbuscular mycorrhizal fungi, with 65, 48, 43, 21 and 5 researches, respectively (Figure 2). In the study of Silva (2016), the authors identified 506 species of plants among all the keywords used that allowed the recovery of 5913 articles searched for in their research. The species that returned the most results were: Brassica sp. (269), Salix sp. (183) and Populus sp. (176). The genus Brassica belongs to the family Brassicaceae and its representatives are widely known for their economic interest in food such as mustard, turnip and cabbage, (Herklotz et al. 2010; Szczygłowska et al., 2011), in addition to having species such as *Brassica juncea*, which has a high level of tolerance to toxic elements. Also highlighted, the genera *Salix* and *Populus*, which belong to the Salicaceae family, are characteristic as trees and shrubs, with potential for phytoremediation and are also used as ornamental plants, in the production of wood, in basketry, and in the restoration of water courses. (Sun et al. 2012).

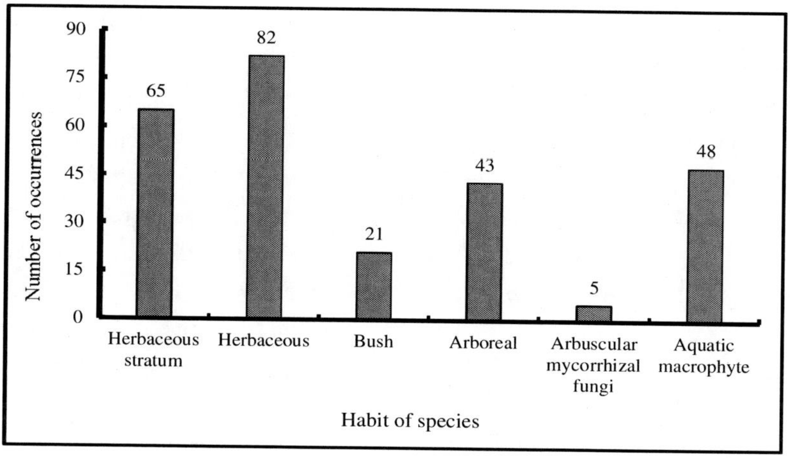

Figure 2. Number of researches published concerning plant groups regarding their tolerance to toxic metals in Scielo database from 2000 to 2020.

In addition, once the studied species were surveyed, they were categorized accordingo to their biome of occurrence. Thus, we identified that species from Cerrado biome (83 occurrences) are the target of most phytoremediation research, followed by Atlantic Forest biome (73 occurrences) (Figure 3).

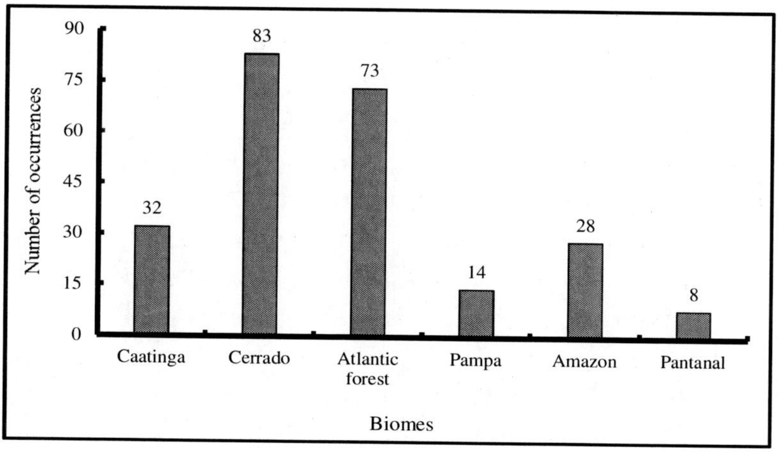

Figure 3. Representation of number of plant species by biome that were studied concerning their tolerance to toxic metals from 2000 to 2020.

When dealing with the greatest occurrences, the Cerrado biome was identified in a study carried out by the Ministry of the Environment (Mma 2011), more than 11,000 species, of which 4,400 are endemic and the heterogeneity of the space is a determining factor for diversity. With the classification of this vast biodiversity, this biome was qualified as the "richest savanna phytophysionomy in the world", in addition to the name "cradle of waters", since it houses the headwaters of the main rivers of Amazon, Prata and São Francisco River (Mma 2011).

However, Da Silva et al. (2014) reported that the amount of waste generated from the urban, industrial or agro-industrial network associated with the high cost of chemical fertilizers is becoming an increasingly attractive option with the use of organic waste in agriculture; and, among the problems found, it is noted that the presence of toxic metals can restrict and limit its use, since it can be absorbed by plant's root system and enter the food chain.

Thus, elements such as barium (Ba), lead (Pb), cadmium (Cd), mercury (Hg), chromium (Cr), manganese (Mn) and nickel (Ni), found in anthropic sources, when disposed in excessive quantities in soil can cause toxic effects in both plants and soil biota (Lequeux et al., 2010; Cambrollé et al., 2015).

Regarding the hyperaccumulator plants, the Cerrado has great potential to be studied as its soils are acidic and rich in metals like iron, manganese and aluminum; in such case the plants are adapted to these conditions in order to reach their complete life cycle (Meyer et al. 2016). Some features of the cerrado biome, the most common are: low level of nutrient availability (N and P), high phosphorus fixation, low cation exchange capacity, high acidity coupled to aluminum and manganese toxicity (Chaves 2001). So, the phytoremediation potential of cerrado's native plants is determined by the characteristics presented, and which include rustic plants that have adapted, differentiated biology and that, under edaphoclimatic conditions, other plants would not survive. Thus, one of the biggest barriers to the development of the phytoremediation technique in the Cerrado today is the lack of studies with native plants (Haridasan, 1994).

The desirable criteria for phytoremediatior plants, considering the characteristics of the soil, Procópio et al. (2009b) consider the high transpiratory capacity as important factors; a high rate of growth and production of biomass and root exudation; deep and dense root system; resistance to pests and diseases; capacity to use atmospheric nitrogen fixation; adaptability to the contaminated site; easy acquisition or multiplication of propagules, as well as easy control or subsequent eradication; high association with mycorrhizal fungi; among others.

The plant species from Cerrado biome are extremely complex in structures and the great variety of shape, size and degree of sclerophilia of the leafy limb of woody species, in addition to the great diversity of life, are outstanding qualities of any area of this environment (Franco, 2005). According to Lamego &Vidal (2007), the advantages of using plants as phytoremediation include high biomass production, rapid growth, vigor, pollution tolerance and competitiveness. However, it is to be expected that the use of plants in this biome as a phytoremediation potential is being underestimated and, consequently, it would be interesting to increase the characterization studies of plants in this biome regarding the phytoremediation potential, as it would mean another advance in increase the value of Brazilian biodiversity.

Searching of Phytoremediation Researches Published in Journals Indexed in Scopus Database

We looked to identify the trend of publications about phytoremediation in Scopus database. There was increase in the number of researches published in journals indexed in Scopus database from 2000 to 2020. In 2020, 1574 publications were recovered (Figure 4), indicating an increase of 91% in relation to the year 2000, and the journals areas with the highest publications were environmental sciences and agricultural and biological sciences, presenting 9696 and 5440 studies, respectively (Figure 5).

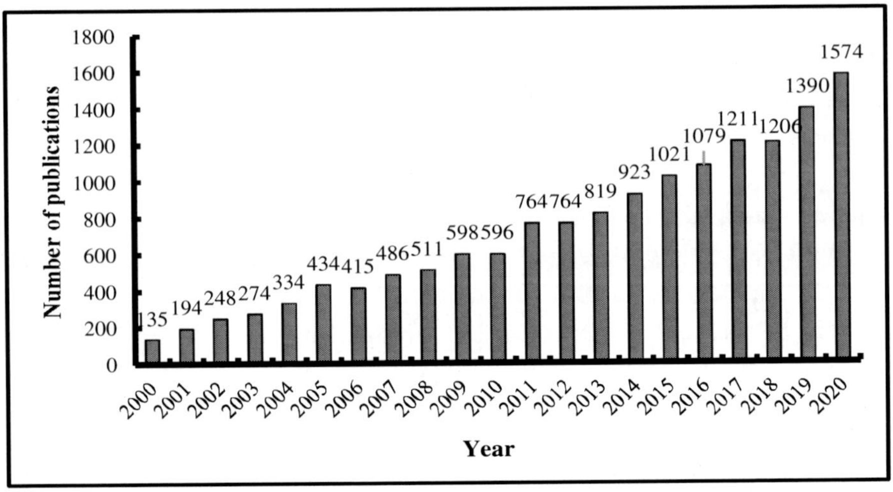

Figure 4. Number of researches published about phytoremediation in journals indexed in Scopus database from 2000 to 2020.

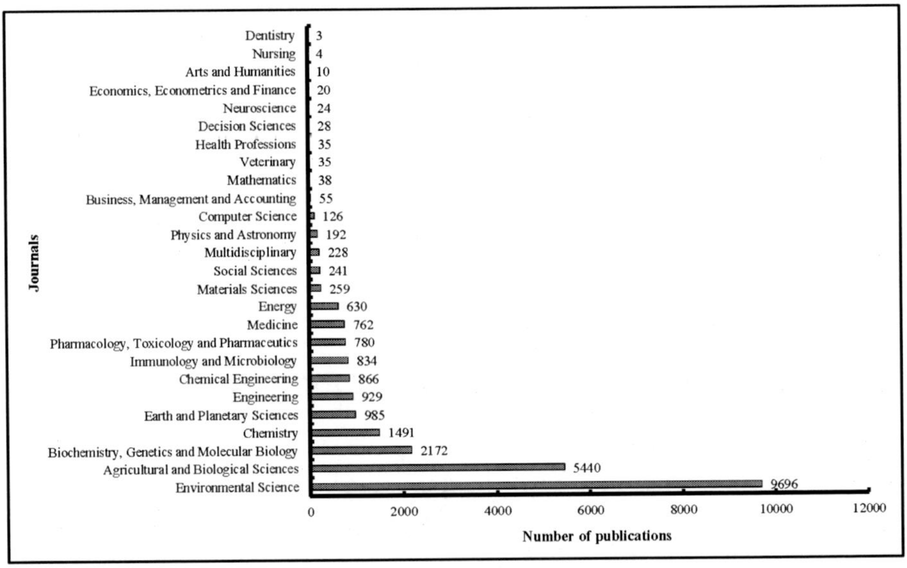

Figure 5. Knowledge areas of journals tht published researches about phytoremediation from 2000 to 2020.

Considering a total of 15,086 publications retrieved from the Scopus database between 2000 and 2020, 11,901 publications were scientific article type, indicating the carry out of original research with a significant increase of 91% in the number of researches published between 2000 (104 published articles) and 2020 (1339 published articles) (Figure 6).

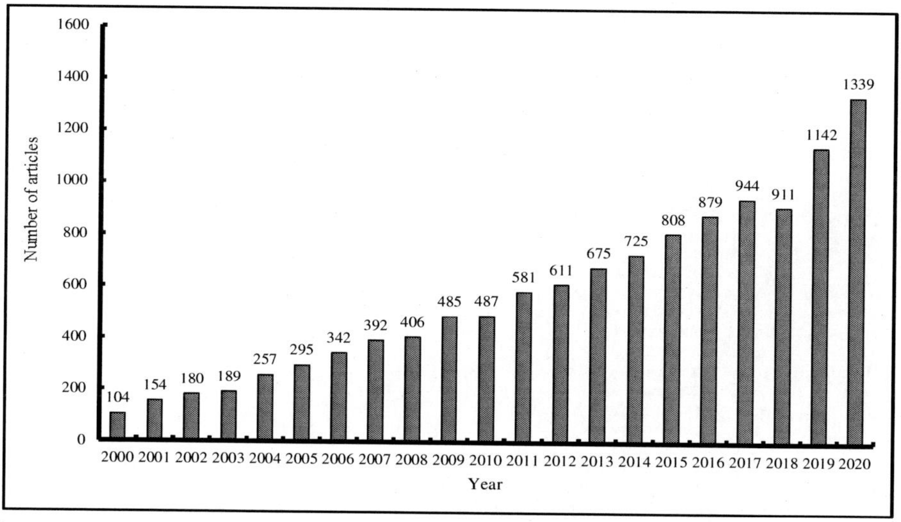

Figure 6. Number of articles published researches about phytoremediation in Scopus database from 2000 to 2020.

The country with the greatest contribution in the phytoremediation area was China, with 3668 articles, followed by the United States of America with 2167 articles and India with 1830 articles (Figure 7). This is a reflection that the Asian continent leads studies involving the technique and this may be related to the high demand for knowledge generation in sustainable technologies for the recovery of contaminated areas. In the study of Silva (2016), it was observed that the USA occupied the first position in number of articles on phytoremediation. For most of knowledge areas, scientific researches is generally centralized in industrialized countries and with greater risks of contamination.

It is possible to concluded that Asia, North America and European countries are dominant in the production of scientific knowledge related to phytoremediation (Pierro, 2014). India and Spain, despite occupying the ninth and tenth position in general scientific production, are in third and fourth position, respectively, in scientific production about phytoremediation (Figure 7), suggesting that these countries have invested in such researches, even with a less number of researchers than the countries shown in fisrt and second position.

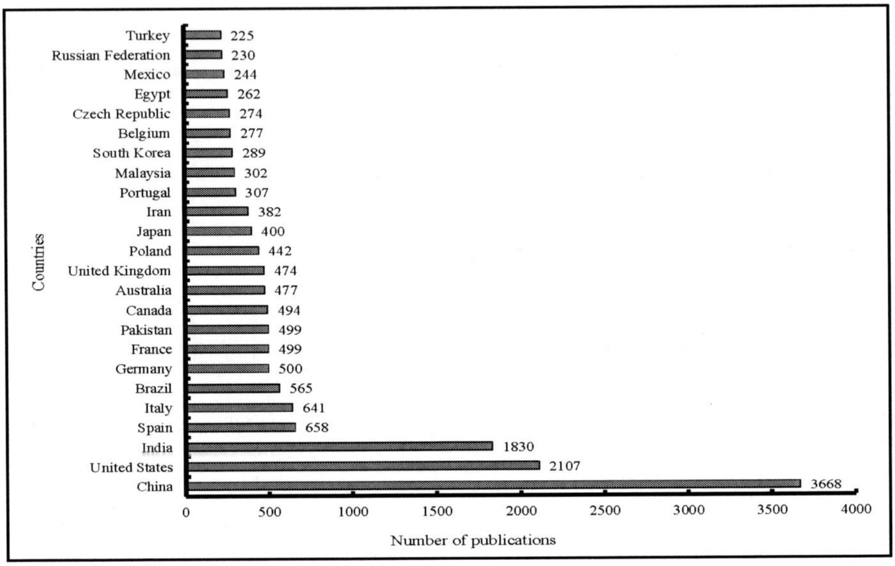

Figure 7. Number of researches published about phytoremediation by country from 2000 to 2020.

The Brazil occupies the sixth position in scientific knowledge about phytoremediation, which can be considered relevant on the world stage. According to Silva (2016), in the ranking of global scientific production, only Brazil, in all Latin America, was highlighted, occupying the fifteenth position and, considering studies on phytoremediation, it was also the only country in Latin America highlighted.

Considering only the publications of Brazilian authors who published abroad, 726 researches were retrieved, which 484 were published as articles. In this scenario, 250 papers were published in the magazine Agriculture in Biological Sciences and 235 papers published in the magazine Enviromental Science (Figure 8). Publications as articles grew 97% from 2000 to 2020 (Figure 9), demonstrating that Brazilian researches have been carried out and published in journals indexed in the Scopus database. This information is very relevant from the point of view of improving the visibility of research carried out in Brazil. In addition, it is possible to conclude that Brazilian researchers prefer to publish in international journals over national journals.

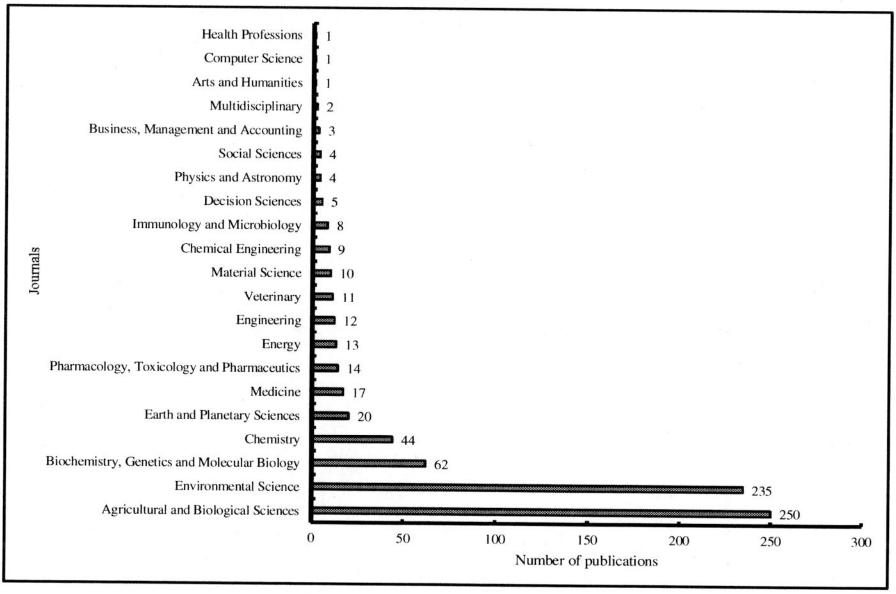

Figure 8. Number of researches of Brazilian researchers published about phytoremediation according to area of knowledge in Scopus database from 2000 to 2020.

By searching for studies published in the worldwid database, with publications by Brazilian authors, this amount increased significantly from the year 2001 (2 publications) compared to the year 2020 with 79 total researches, 53 of which with inorganic compounds and 26 with organic, reaching the peak of publications (Figure 9 and 10).

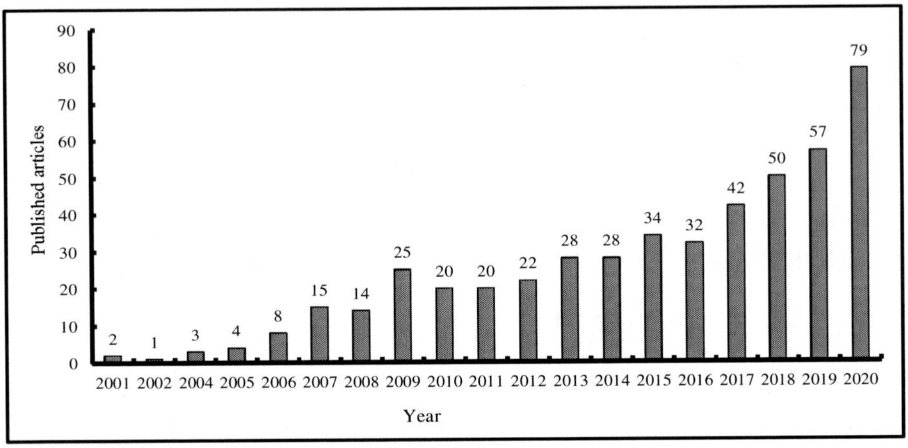

Figure 9. Number of researches of Brazilian researchers about phytoremediation published in journals indexed in Scopus database from 2000 to 2020.

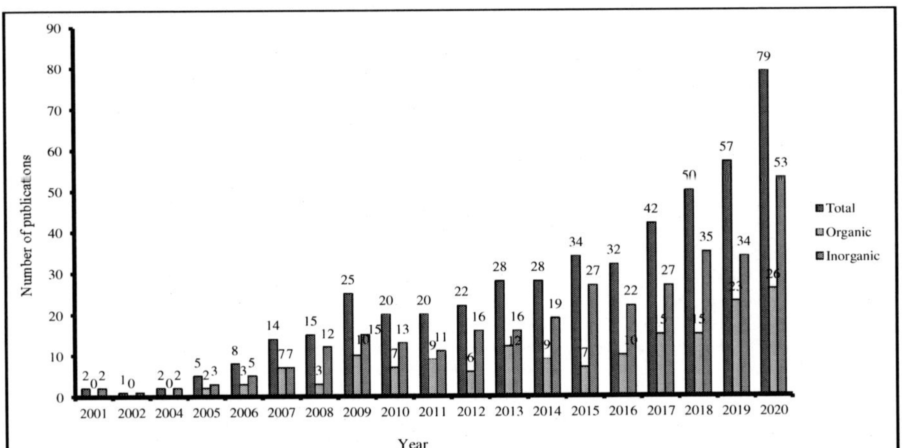

Figure 10. Number of researches of Brazilian researchers regarding organic and inorganic pollutants that were published in Scopus database from 2000 to 2020.

Considering the total researches published by Brazilian researchers, it was searched which journals indexed in the Scopus database published on Phytoremediation. In the period from 2000 to 2020, the Journals "International Journal of Phytoremediation" and "Water, Air and Soil Pollution" published a total of 31 and 25 researches, respectively (Table 3).

For both journals, the largest number of publications occurred in journals that studied inorganic contaminants (18 papers published in the Water, Air and Soil Pollution and 20 in the International Journal of Phytoremediation). Thus, the same worldwide scenario was found when compared to publications in Brazil. In this scenario, it is evident that there is a greater concern in solving problems related to contamination by inorganic than organic compounds, or that there are, in fact, greater problems in reality with inorganic contaminants due to industrial and agricultural activities.

Table 3. Total number of researches published in the first 15 journals indexed in Scopus database as well as the number of researches with organic and inorganic pollutants from 2000 to 2020

Journals	Total	Inorganic	Organic
International Journal of Phytoremediation	31	20	11
Water, Air, and Soil Pollution	25	18	7
Chemosphere	23	19	4
Planta Daninha	20	5	15
Environmental Science and Pollution Research	18	13	5
Revista Brasileira de Ciência do Solo	18	14	4
Revista Brasileira de Engenharia Agrícola e Ambiental	13	8	5
Ecotoxicology and Environmental Safety	9	7	2
Environmental Pollution	9	7	2
Scientia Agricola	9	4	5
Bragantia	6	3	3
Ecological Engineering	6	2	4
Journal of Soils and Sediments	6	5	1
Journal of Hazardous Materials	6	5	1
Brazilian Journal of Biology	6	6	0
Environmental and Experimental Botany	6	5	1
Revista Árvore	6	4	2
Revista Ciência Agronômica	6	3	3
Pesquisa Agropecuária Tropical	5	2	3

We've carried out the classification of journals indexed in Scopus according to impact factor. Most of the publications are in the IF range of 0-0.99, with 63 researches about organic pollutants and 112 researches with inorganic pollutants (Table 4). Considering the higher IF, from 7-7.99, there were only 1 publication considering organic pollutant (Table 4).

Table 4. Number of studies with organic and inorganic pollutants in scientific journals indexed in Scopus data base and organized according to impact factors

Impact Factor		
	Organic	**Inorganic**
0-0.99	63	112
1 – 1.99	48	89
2 – 2.99	17	51
3 – 3.99	10	20
4 – 4.99	14	32
5- 5.99	8	10
6 – 6.99	2	7
7 – 7.99	1	0

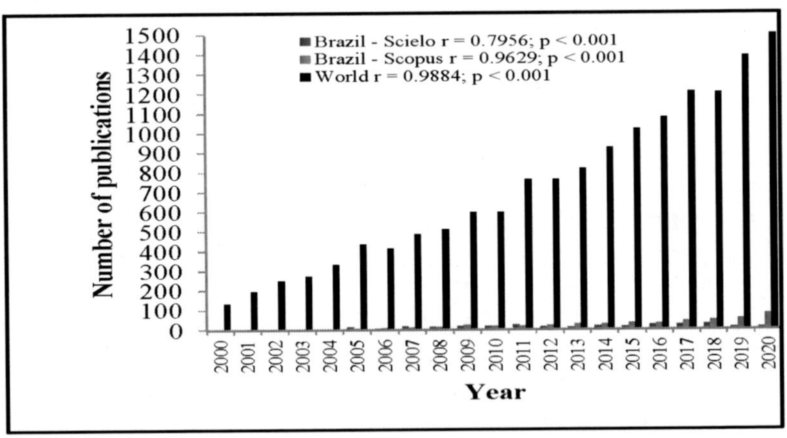

Figure 11. Time course of researches about phytoremediation of Brazilian reasearchers in Scielo and Scopus database from 2000 to 2020.

In order to identify the trend in publications about phytoremediation in Scielo and Scopus database we ploted the trend together and it is clear that while there is a global trend of significant increase in publication in 20 years ($r = 0.96$; $P < 0.001$), at the national level, this trend has decreased ($r = 0.79$; $P < 0.001$) (Figure 11).

As shown in Figure 12, in 2020 we observerd the largest contribution (5.02%) of researches by Brazilian authors published in relation to researches published in the world. It is observed the dimension of the evolution of studies carried out with this focus and that, although the Brazilian scenario is much lower than expected when compared to the World, there is a trend in a greater number of publications.

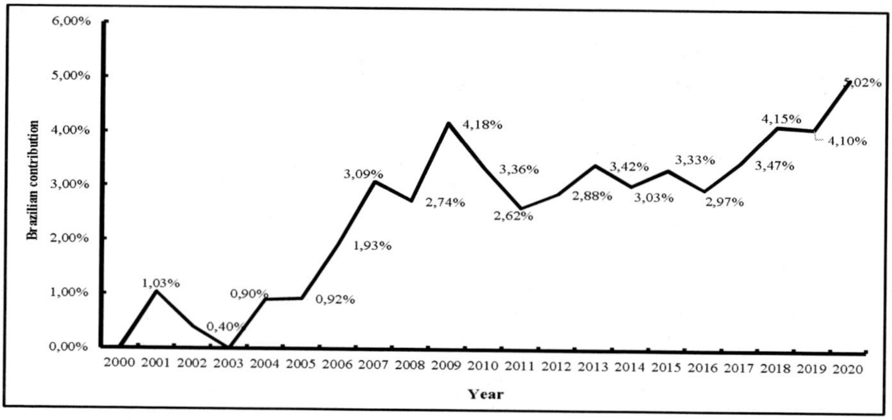

Figure 12. Brazilian contribution to phytoremediation researches that were published in journals indexed in Scopus database from 2000 to 2020.

CONCLUSION

Most of researches focus to publish in journals indexed in Scopus than Scielo. Brazilian contribuition to phytoremediation is still low although the high biodiversity and potential to discover new hyperaccumulator species as well as appropriate species to phytostabilization and phytoextraction.

ACKNOWLEDGMENTS

The authors thanks to CNPq, CAPES, FAPEG, FAPESP and IF Goiano for all support during this research.

REFERENCES

Al Naggar, Y. A., Naiem, E. A., Seif, A. I., Mona, M. H. (2013). "Honey bees and their products as a bio-indicator of environmental pollution with heavy metals." *Mellifera*, 13, 10–20.

Ali, H. et al. (2013). "Phytoremediation of heavy metals concepts and applications. " *Chemosphere,* 91, 869-881.

Brauna, A. B., Trentin, A. W., Balestrinc, D., & Thoméd, A. (2017). "Remediação sustentável: um panorama geral das publicações".In 2º *Congresso de engenharia ambiental do Sul do Brasil.* ["Sustainable remediation: an overview of publications" In *2nd Environmental Engineering Congress in Southern Brazil*]

Cambrollé, J., García, J. L., Figueroa, M. E., Cantos, M. (2015). "Evaluating wild grapevine tolerance to copper toxicity". *Chemosphere*, 120, 171-178.

Chaves, L. J. "Melhoramento e Conservação de Espécies Frutíferas do Cerrado". (2001). In *Anais do 1º Congresso Brasileiro de Melhoramento de Plantas. Goiânia. Sociedade Brasileira de Melhoramento de Plantas*, 7. ["Improvement and Conservation of Cerrado Fruit Species". (2001). In *Annals of the 1st Brazilian Plant Breeding Congress. Goiânia. Brazilian Society of Plant Breeding*]

Chen, X. W., Tsz-Fung Wong, J., Mo, W. Y., Man, Y. B., Wang-Wai Ng, C., Wong, M. H. (2015). "Ecological Performance of the Restored South East New Territories Landfill in Hong Kong (2000–2012)", *Land Degrad*, 1,1–13.

Chibuike, G. U., Obiora, S. C. (2014). "Heavy Metal Polluted Soils: Effect on plants and bioremediation methods", *Applied and environmental soil science*, 12, 2014.

Da Silva, M. A., Büll, L. T., Miggiolaro, A. E., Antonangelo, J. A., Muniz, A. S. (2014). "Fitodisponibilidade de metais utilizando ácidos orgânicos após sucessiva aplicação de resíduos no solo". *Revista Brasileira de Engenharia Agrícola e Ambiental,* 18, 1287-1295. [Phytodavailability of metals using organic acids after successive application of residues in the soil. *Brazilian Journal of Agricultural and Environmental Engineering*]

Franco, A. C. (2005). "Biodiversidade de forma e função: implicações ecofisiológicas das estratégias de utilização de água e luz em plantas lenhosas do Cerrado". *Cerrado ecologia, biodiversidade e conservação,* 179- 196. ["Biodiversity of form and function: ecophysiological implications of strategies for using water and light in woody plants in the Cerrado". *Cerrado ecology, biodiversity and conservation*]

Ghoneim, A. M., Al-Zahrani, S., El-Maghraby, S., Al-Farraj, A. (2014). "Heavy metal distribution in Fagonia indica and Cenchrus ciliaris native vegetation plant species", *Journal Food Agric Environmental,* 12, 320–324.

Haridasan, M. (1994). Solos. In *Cerrado - Caracterização, Ocupação e Perspectivas.* (M. Novaes-Pinto, ed.). Editora Universidade de Brasília, 321-344. [Soils. In *Cerrado - Characterization, Occupation and Perspectives.* (M. Novaes-Pinto, ed.). Publisher University of Brasilia]

Heiderscheidt, D., Pereira, J., Burghardt, J. E., da Silva, L. A., de Oliveira, S. C. (2016). "Conceitos aplicados à poluição do solo decorrente do derrame de petróleo e seus derivados." *Maiêutica-Gestão Ambiental*, 4, 1. [Concepts applied to soil pollution resulting from the spill of oil and its derivatives. *Maiêutica-Environmental Management*]

Herklotz, P. A., Gurung, P., Heuvel, B. V., Kinney, C. A. (2010). "Uptake of human pharmaceuticals by plants grown under hydroponic conditions. " *Chemosphere*, 78, 1416-1421.

Lamego, F. P., Vidal, R. A. (2007). Fitorremediação: plantas como agentes de despoluição? *Pesticidas ecotoxicol. e meio ambiente*, 17, 9-18. [Phytoremediation: plants as depollution agents? *Ecotoxicol pesticides. and environment*]

Lequeux, H., Hermans, C., Lutts, S., Verbruggen, N. (2010). "Response to copper excessin Arabidopsis thaliana: impact on the root system architecture, hormone distribution, lignin accumulation and mineral profile. " *Plant Physiology and Biochemistry*, 48, 673-682.

Li, X. L., Gao, J., Brierley, G., Qiao, Y. M., Zhang, J., Yang, Y. W. (2013). "Rangeland degradation on the Qinghai–Tibet Plateau: implications for rehabilitation." *Land Degrad*, 24, 72–80.

Meyer, S. T., Castro, S. R., Fernandes, M. M., Soares, A. C., Souza, G. A. F., Ribeiro, E. (2016). "Heavy-metal-contaminated industrial soil: Uptake assessment in native plant species from Brazilian Cerrado." *International journal of Phytoremediation,* 18, 832-838.

Morgan, R. (2013). "Soil, heavy metals, and human health." *Soils and human health*, 59–82,

Ministério do Meio Ambiente (2011). *Guia de campo: vegetação do Cerrado 500 espécies.* Brasília: MMA/SBF, 532. Translation: [Ministry of Environment (2011). A field guide: 500 species of "*cerrado*" vegetation.] Brasília: MMA/SBF, 532.

Moosavi, S. G., Seghatoleslami, M. J. (2013). "Phytoremediation: A Review. " *Advance in Agriculture and Biology*, 1, 5-11.

Pierro, B. (2014). "Ciência em evolução. " *Pesquisa FAPESP*, 222, 32–35. [Science in evolution. *FAPESP Research*]

Pimentel, V. P., Vieira, V. A. M., Mitidieri, T. L., Oliveira, F. F. S., Pieroni, J. P. (2015). "Biodiversidade brasileira como fonte da inovação farmacêutica: uma nova esperança? " *Revista do BNDES*, 43, 41-89. [Brazilian biodiversity as a source of pharmaceutical innovation: a new hope? *BNDES Magazine*]

Procópio, S. O., Pires, F. R., Santos, J. B., Silva, A. A. (2009b). Fitorremediação de solos com resíduos de herbicidas. *Aracaju: Embrapa,* 32. [Phytoremediation of soils with herbicide residues. *Aracaju: Embrapa*]

Silva, E. C. Uma síntese dos estudos com fitorremediação nos últimos anos (1991-2014). (2016). *Universidade Estadual de Goiás.* [A summary of studies with phytoremediation in recent years (1991-2014). (2016). *State University of Goiás.*]

Souza, E. P. D., Silva, I. D. F. D., Ferreira, L. E. (2013). "Mecanismos de tolerância a estresses por metais pesados em plantas." *Current Agricultural Science and Technology*, 17, 167. [Mechanisms of stress tolerance for heavy metals in plants.]

Sun, X. C., Wei, H., Xie, X. H., Jia, Z. M., Meng, X. F. (2012). "Bioaccumulation and photosynthesis response of Salix variegata to cadmium under hydroponic culture." *Environmental Science*, 25, 220 – 225.

Szczygłowska, M., Piekarska, A., Konieczka, P., Namieśnik, J. (2011). "Use of Brassica plants in the phytoremediation and biofumigation processes." *International Journal Molecular Science*, 12, 7760-7771.

INDEX

A

B

C

D

E

F

G

H

I

L

M

N

O

P

R

S

T

U

V

W

X

Z